Titanium Alloys – Anticorrosion and **Etching**

by

Cui Lin and Meifeng Wang

Canamaple Academia Services
Camdemia   Halifax, Canada

This book is to provide research advances focusing on anticorrosion performance, surface treatment and chemical etching of titanium alloys. It can serve as a reference book for academics, researchers, postgraduate students pursuing materials and corrosion engineering as subject of study and research. It may also be used by engineers and professionals working on application technology development, engineering design and processing of titanium material.

ISBN 978-0-9948791-2-7 (print)
ISBN 978-0-9948791-4-1 (ebook)

Titanium Alloys – Anticorrosion and Etching
Cui Lin & Meifeng Wang

Subject headings: titanium alloy / corrosion and protection / materials engineering

Collection in Library and Archives Canada

Published and distributed by
Canamaple Academia Services
Halifax, Canada, B3M 2Y2
http://press.camdemia.ca, press@camdemia.ca

First Printing, February, 2017.

Disclaimer:

While all care is taken to ensure the accuracy and correctness of the information contained in this textbook, the author and the publisher shall not be liable for any damage to property or persons as a result of using the information contained herein.

To My Parents

Cui Lin

Table of Contents

Preface

Titanium alloys developed in the 1950s have become a new type of structural material due to its excellent comprehensive performance. Recently, a boom in the global titanium industry and titanium processing technology has opened up a wide variety of applications for titanium alloys. The corrosion, friction and wear properties of titanium products in various environments play a significant role in affecting their operational reliability and service life. Therefore, it is vital to study the corrosion behavior of titanium materials and develop surface treatment methods to enhance their surface properties. This is becoming increasingly important in improving the reliability, safety and durability of titanium products and reducing the loss of titanium resources. Because of their low thermal conductivity, low elastic modulus and chemical reactions with tools, it is difficult to manufacture titanium alloys by conventional mechanical machining. Chemical etching, which is a controlled corrosion process, is considered an alternative processing method. This has necessitated the need for exploring its technology and etching mechanism for use in titanium alloys.

The current published books on titanium alloys mainly involve metallurgy, processing and applications. There are few books containing detailed information about corrosion and protection of titanium alloys. My research work on titanium alloys started ten years ago. This motivated me to write this book.

In this book, the basic knowledge of titanium and titanium alloys is briefly introduced, the corrosion properties and corrosion data as well as surface treatment methods are summarized, and the work on cavitation erosion behavior, chemical etching and electroless plating of titanium alloys from my research group are emphatically elaborated.

Chapter 1 introduces history, basic properties, class, present reserve and usage amount in various fields of titanium and titanium alloys.

Chapter 2 focuses on anticorrosion performances in a variety of environments and provides some data, the sources of which are from literature and handbooks of non-ferrous materials. On the basis of this, the suggestions for material selection, practical use design and some preventive measures are proposed.

Chapter 3 gives a review of methodologies and influencing factors of cavitation erosion, and presents the research on the cavitation erosion behavior of titanium alloys in the working fluid of aqueous lithium bromide solution.

Chapter 4 provides an overview of history, characteristic, clasification and application of chemical etching. The study on the process and mechanism of chemical etching applied for titanium alloys is mainly discussed. It involves the influencing factors, the effect of chemical etching on the titanium substrate properties, the corrosion dissolution mechanism and kinetics.

Chapter 5 contains a summary of surface treatment techniques for titanium alloys, and the research on the preparation of electroless plating Ni-P and Ni-P-MoS$_2$ coating on titanium alloys, the increase of coating adhesion and investigation of wear resistance of the coatings.

This book is meant to present up-to-date experimental data and research in the fields of corrosion, chemical etching and surface treatment of titanium alloys. I hope it can provide useful fundamental and practical information for academics, researchers, postgraduate students, engineers and professionals interested in the sustainable development of titanium materials and the design, processing and utilization of titanium components in industrial equipment.

Acknowledgement

I would like to express my gratitude to my graduate students and colleagues, who made their important contributions to the research covered from Chapter 3 to Chapter 5 in this book and provided assistance in taking some pictures and making some figures. I would also like to thank coauthor Dr. Meifeng Wang for his work and ideas in writing this book. I sincerely appreciate Prof. Steve Zou for his encouragement, constructive suggestions in writing and help in editing, as well as Lisa K. Zou for taking her time to read and making helpful comments for improvements. Finally, I want to thank my family for their support to complete this book.

Cui Lin
January, 2017

Symbols and Abbreviations

Below is the list of some symbols and abbreviations used in this book.

Symbols

T	temperature
λ	coefficient of thermal conductivity
R	resistivity
E	electric potential
E_{corr}	corrosion potential
i_{corr}	corrosion current density
E_p	maintaining passive potential
i_p	maintaining passive current density
E_{tp}	transpassive potential
R_p	polarization resistance
HV	Vickers hardness
HK	Knoop hardness
S_c/S_A	area ratio of cathode to anode
p_d	pulse duration
p_t	pulse times
i	current density

Abbreviations

CP-Ti	commercially pure titanium
hcp	hexagonal close-packed
bcc	body-centered cubic
fcc	face-centered cubic
SCE	saturated calomel electrode
SHE	standard hydrogen electrode
OCP	open circuit potential

Chapter 1 - Introduction

1.1 History of Titanium

The metal element titanium (Ti) has a wide distribution in the Earth's crust, where it is present with a content of about 0.63% and is the tenth most abundant element after oxygen, silicon, aluminum, iron, calcium, sodium, potassium, magnesium and hydrogen. Because of its strong binding capacity with oxygen, titanium mainly exists in the form of oxide in nature, and is rarely found in pure form. There are more than 100 types of titanium-containing minerals that have been found. Of these mineral deposits, the most important sources of titanium are rutile and ilmenite, and the secondary are leucosphenite, anatase, brookite and perovskite.

In 1791, a new element was first discovered in dark, magnetic iron sand (ilmenite) by William Gregor, a British chemist and amateur mineralogist. In 1795, this unknown element was identified from rutile by Martin Heinrich Klaproth, a German chemist. He named this element titanium after the Titans of Greek mythology. From the end of the 18th century to the early 19th century, different methods were tried to extract titanium from minerals. But it was not until 1910 that titanium, which was deformable at high temperatures and contained low content oxygen, was produced through reacting titanium tetrachloride ($TiCl_4$) with sodium in laboratory by Matthew A. Hunter in the United States. Pure titanium was obtained via magnesium reduction of $TiCl_4$ by Metallurgist William J. Kroll in Luxembourg in 1940. Nowadays the Hunter process and the Kroll process have developed into the dominant methods for industrial production of titanium. In the 1950s, the production of titanium began industrialization.

Some chemical elements, such as aluminium, vanadium, molybdenum, manganese and iron, etc, can be added in titanium to produce titanium alloys. The first practical titanium alloy is Ti-6Al-4V, successfully made in 1954 in the United States. Owing to its good properties in heat resistance, strength, plasticity, toughness, formability, weldability, corrosion resistance and biocompatibility, it is the most commonly used today and accounts for 75%-85% of all titanium alloys. In the 1950s and 1960s major development focused on high temperature and structural titanium alloys, which are used for aero-engine and airframe structures, respectively. A number of corrosion resistant titanium alloys were exploited in the 1970s. Since the 1980s, corrosion resistant and high strength titanium alloys were further developed. The temperature of heat resistant titanium alloy was upgraded from 400°C in the 1950s to 600-650°C in the 1990s. The emergence of Ti_3Al and TiAl based alloys has advanced titanium alloy's use from the cold end to the hot end of an engine.

The article "Titanium-the infinite choice" by International Titanium Association (2011) and the book "Titanium and Titanium Alloys-fundamentals and application" by Leyens and Peters (2003) give a more detailed description of the history of titanium and the development of titanium materials.

1.2 Crystal Structure and Classification of Titanium Alloys

Titanium is a transition metal having the appearance of silver metallic luster. Titanium element is positioned in period 4, group 4 in the periodic table of elements. Its atomic number is 22 and relative atomic mass is 47.90. The nucleus with the radius of 147 pm consists of 22 protons and 20-32 neutrons. The configuration of 22 outer shell electrons is $1s^22s^22p^63s^23p^63d^24s^2$. The ability of an atom to lose an electron is measured by ionization energy. The 4s and 3d electrons have ionization energy of less than 50 eV and are easy to lose, while 3p electrons have ionization energy above 100 eV and are hard to remove. Hence, valence electrons of the titanium atom are $4s^23d^2$. The highest oxidation state of titanium is usually positive tetravalent (+4) (RSC 2016). Table 1-1 lists the atomic data of titanium.

	22	47.90
Ti		
Titanium		

Table 1-1 Atomic data of titanium

In the periodic table of elements	Group	4
	Period	4
	Atomic number	22
	Relative atomic mass	47.90
	Atomic radius	147 pm
	Electron configuration	$1s^22s^22p^63s^23p^63d^24s^2$
	Element category	transition metal
Ionization energy (J)	1st 4s	1.09×10^{-18}
	2nd 4s	2.17×10^{-18}
	3rd 3d	4.40×10^{-18}
	4th 3d	7.06×10^{-18}
	5th 3p	16.06×10^{-18}
	6th 3p	19.51×10^{-18}

There are two allotropes for pure titanium (Wikipedia-titanium 2017). One is α Ti with a hexagonal close-packed crystal (hcp) structure which is stably present below 882.5°C, and the other is β Ti with a body-centered cubic crystal (bcc) structure existing between 882.5°C and the melting point. Figure 1-1 illustrates the unit cell of α Ti and β Ti. For α Ti, the lattice parameters at 20°C are a=0.2950 nm, c=0.4683 nm and c/a=1.587. Regarding β Ti, the lattice parameters at 20°C and 900°C are a=0.3282 nm and a=0.3306 nm, respectively.

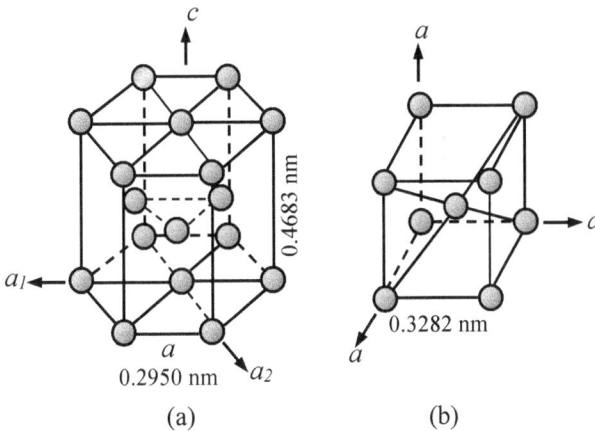

Figure 1-1 Unit cell of α Ti (a) and β Ti (b)

Titanium alloys are generally classified into three main groups (Wikipedia-Titanium alloy 2017). Figure 1-2 schematically demonstrates the classification of titanium alloys.

1) Commercially pure titanium (CP-Ti) and alpha (α)/near alpha titanium alloys

This group can be divided into:

- CP-Ti
- α-Ti
- Near α-Ti

The microstructure in this group is hexagonal α phase. The alloys possess medium strength, good corrosion resistance, and good creep strength at high temperatures. They are normally non-heat treatable but weldable.

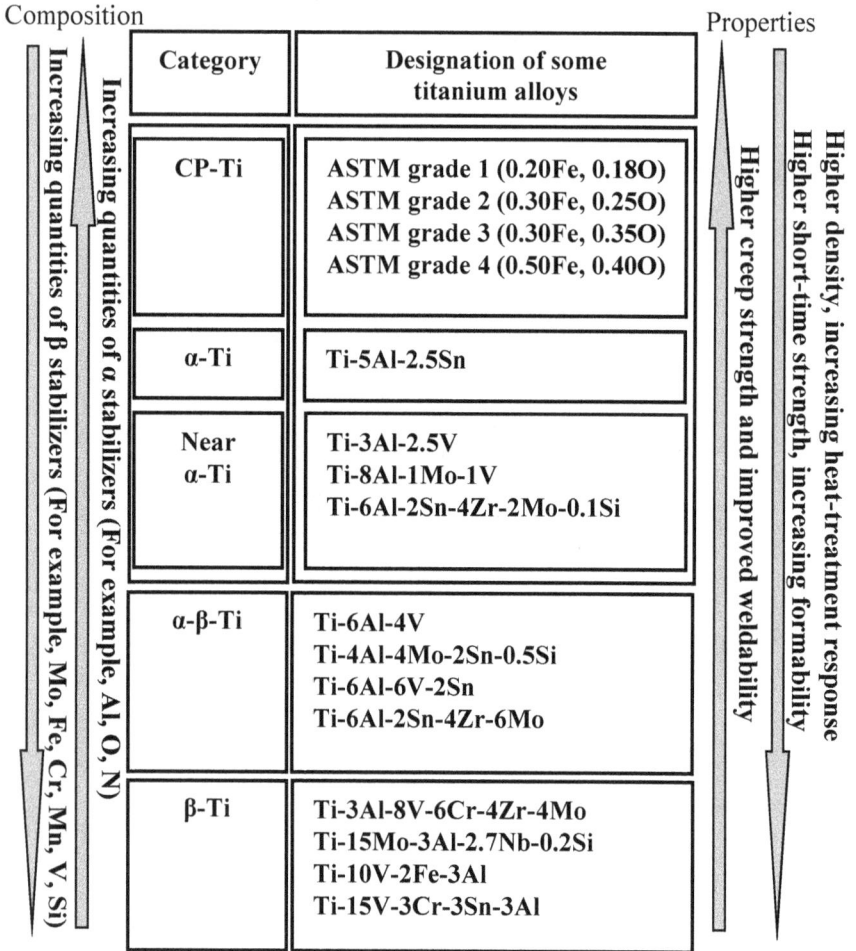

Figure 1-2 Schematic diagram of classification of titanium alloys

CP-Ti contains Fe and interstitial elements (O, C, or N) which are soluble in α phase. The grade is determined by O content and the strength is affected by O and N content.

In α-Ti alloys, Al and O are the main alloying elements, which provide solid solution strengthening. High amounts of aluminum can contribute to oxidation resistance at high temperatures. Neutral alloying elements (such as Sn) are helpful to improve ductility.

For near α-Ti alloys, besides alpha-phase stabilizers, small amounts of β stabilizers (Mo, V, Si) are added, giving a microstructure of β phase dispersed in the α phase structure.

2) Alpha-beta alloys (α-β-Ti)

α-β-Ti alloys include some combination of both alpha and beta stabilizers. The alloys in this group are metastable and can be heat treated. Compared to α-Ti alloys, α-β-Ti alloys show enhanced strength and formability. Ti-6Al-4V is the typical and widely used alloy in this group (Gurrappa 2003).

3) Beta alloys (β-Ti)

β-Ti alloys have sufficient beta stabilizers. After quenched, solution treated and aged, the strength of the alloys can be improved. The alloys are metastable and possess a bcc crystal structure.

1.3 Basic Properties of Titanium Alloys

This section provides an overview of physical properties, including thermal, electrical, magnetic and mechanical properties, and chemical properties of titanium alloys (Donachie 2000, Total Materia 2005, Lütjering and Williams 2007).

Titanium alloys feature low density, high strength, excellent corrosion resistance, good heat resistance, wide service temperature range, hydrogen storage capacity, superconductivity, high damping, shape memory and superelasticity (Figure 1-3). Presently, they have become the first choice for structural materials, newly functional materials and important biological materials.

1) High strength-to-density ratio

The density of titanium alloys is only about 60% of that of steel, but the strength is 18% higher than that of steel. The specific strength (strength/density) of titanium alloys is much larger than that of other structural metallic materials. It can be produced to components and parts with high strength, good rigidity and light weight.

2) Good high-temperature strength

Titanium alloys can be used in temperatures hundreds of degrees higher than that of aluminum alloys. Titanium alloys have very high strength at temperatures of 150°C to 500°C. Comparatively, the strength of aluminum alloys decreases significantly at 150°C. The operating temperature of titanium alloys can reach 500°C, while aluminum alloys are below 200°C.

3) Superior corrosion resistance

Titanium alloys have excellent corrosion resistance in some aggressive environments, including alkaline solutions, most organic acid solutions, inorganic salt solutions and oxidation media, as a result of the formation of a stable, adherent oxide layer on the surface.

4) Good low-temperature performance

The mechanical properties of titanium alloys can be maintained at low and super low temperatures. At -253°C, titanium alloys still have good ductility and toughness, and cold shortness would not occur. Therefore, they are ideal materials for cryogenic containers, such as liquid fuel storage tanks.

5) Special features

Some titanium alloys containing niobium, nickel and iron have special physical and biological functions. For example, Ni-Ti and Nb-Ti are superconducting alloys, Ti-Ni based alloy is a shape memory alloy and Ti-Fe is a hydrogen storage alloy.

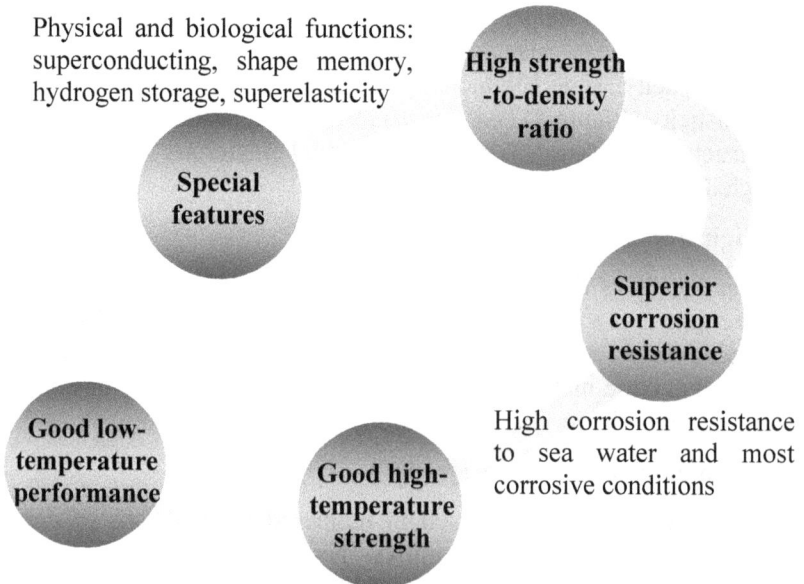

Figure 1-3 Advantages of titanium alloys

1.3.1 Density and Strength-to-Density Ratio

Titanium is a low-density metal. Table 1-2 gives the density at different temperatures. The density and strength-to-density ratio of several metallic materials are compared in Figure 1-4. The density of titanium is approximately 57% of iron.

Aluminum, magnesium and titanium belong to lightweight materials. Among them, titanium has the highest strength-to-density ratio.

Table 1-2 Density of titanium at different temperatures

Temperature (°C)	Density (g/cm³)
20	4.51
870	4.35
900	4.32

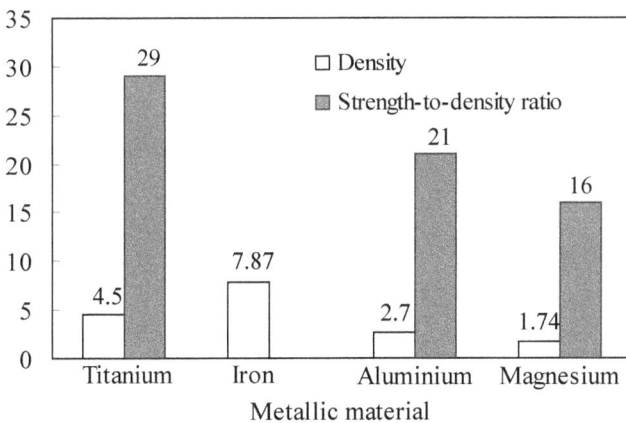

Figure 1-4 Density and strength-to-density ratio of several metallic materials

1.3.2 Thermal Properties

Titanium is a refractory metal with a melting point of 1668±5°C. It is difficult to measure the latent heat of fusion of titanium, because molten titanium has high reactivity with other refractory materials. The latent heat of fusion of titanium which has been obtained currently by measurement ranges between 15.46 and 20.9 kJ/mol. Surface tension of liquid titanium at the melting point is 1.588 N/m and dynamic viscosity of liquid titanium at 1730°C is 8.9×10⁻⁵ m²/s. At atmospheric pressure, the boiling point is 3260±20°C and the latent heat of vaporization is 428.5-470.3 kJ/mol. The critical temperature and pressure are 4350°C and 113 MPa, respectively.

The coefficient of thermal conductivity of CP-Ti increases steadily as the temperature varies from 0 (-273°C) to 250 K (-23°C) until it peaks at 17.56 W·m⁻¹·K⁻¹ at 250 K, which is followed by a slight fall. It decreases to a minimum of 16.89 W·m⁻¹·K⁻¹ at about 580 K (307°C). The coefficient of thermal conductivity of CP-Ti (T>273 K, unit of T is Kelvin) can be calculated by Eq.(1.1).

$$\lambda = 17.56 - 4.6 \times 10^{-3}T + 8.23 \times 10^{-5}T^2$$
$$- 14.7 \times 10^{-6}T^3 + 4.8 \times 10^{-12}T^4 \tag{1.1}$$

The coefficient of thermal conductivity of titanium alloys decreases by about 50% compared with that of pure titanium.

Figure 1-5 shows the coefficient of thermal conductivity of several metallic materials. It can be seen that the thermal conductivity of titanium is about 1/13 of that of aluminum, 1/5 of that of steel, and 1/15 of that of copper.

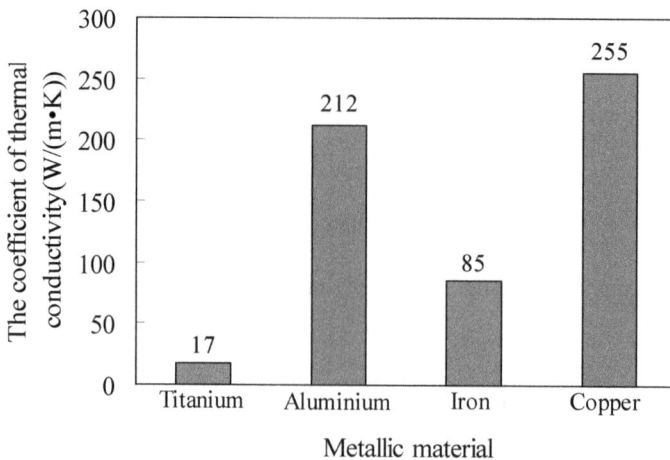

Figure 1-5 Coefficient of thermal conductivity of several metallic materials

The thermal properties of titanium are depicted in Table 1-3.

Table 1-3 Thermal properties of titanium

Property	Value
Melting point	1668±5°C
Boiling point	3260±20°C
Specific heat (at 300 K)	566 J·kg^{-1}·K^{-1}
Latent heat of fusion	5.46-20.9 kJ/mol
Latent heat of vaporization	428.5-470.3 kJ/mol
Coefficient of linear thermal expansion (at 293-373 K)	20-100 ($\times 10^{-6}$K^{-1})
Coefficient of thermal conductivity (at 300 K)	17 W·m^{-1}·K^{-1}

1.3.3 Electrical Properties

Titanium is not a good conductor of electricity, like stainless steel. Its electrical behavior is influenced by the atomic structure. If the conductivity of copper is 100%, titanium has a conductivity of 3.1% by comparison. Impurities in titanium, structure, and temperature can affect the electrical resistance. Due to the presence of impurities, such as oxygen, nitrogen, carbon, iron, etc, the resistance measurement often gives scattered data. When the transition from α phase to β phase occurs, the specific resistance decreases. There is a rise of specific resistance as temperature increases. When the temperature approaches absolute zero, titanium has superconductivity.

The resistivity of CP-Ti is 0.556 μΩ·m at 20°C. The relationship between the resistivity R (μΩ·m) and temperature can be described by

$$R = 0.51 + 2.25 \times 10^{-3}\,T - 8.6 \times 10^{-10}\,T^3 \tag{1.2}$$

For Eq. (1.2), unit of T is °C.

1.3.4 Magnetic Properties

Titanium is slightly paramagnetic. The increase of temperature can improve the susceptibility. The susceptibility of α-Ti at 20°C and β-Ti at 900°C is (3.2±0.4)$\times 10^{-6}$ and 4.5$\times 10^{-6}$, respectively. The magnetic permeability of CP-Ti is 1.0004 H/m (henries per meter).

1.3.5 Mechanical Properties

Titanium has the advantages of both steel (high strength) and aluminum (lightweight). The yield strength of titanium is close to its tensile strength, and the elastic modulus of titanium is small, about 54% of iron.

Generally, α-Ti is more ductile than β-Ti, due to the larger number of slip planes in the bcc structure of the β-phase in comparison to the hcp α-phase. α-β-Ti has a mechanical property which is in between both.

1) Tensile strength

Good mechanical properties of titanium can be obtained by strict control of the appropriate content of impurities and addition of alloy elements.

A rise in the content of impurities, such as oxygen, nitrogen, carbon, et al, can increase the strength of titanium, but at the same time reduce its ductility. The order influenced by the impurities is nitrogen > carbon > oxygen. The tensile strength of titanium and its alloys at ambient temperature ranges from 240 MPa for the softest grade of CP-Ti to more than 1300 MPa for very high strength alloys (Table 1-4). Figure 1-6 shows the comparison of tensile and yield strength of different metals.

2) Elastic modulus and Poisson's ratio

The elastic property of titanium is affected by the working process used to produce the titanium material and intrinsically anisotropic character. Values of elastic modulus typically range from 80 to 125 GPa. Table 1-5 gives the elastic modulus of several titanium materials. Poisson's ratio varies from 0.287 to 0.391 for annealed Ti-6Al-4V.

Table 1-4 Mechanical properties of titanium alloys

Type of titanium		Tensile strength (MPa)	Yield strength (MPa)
CP-Ti	ASTM Grade 1	241	172
	ASTM Grade 2	345	276
	ASTM Grade 3	448	379
	ASTM Grade 4	552	483
α and near α-Ti	Ti-5Al-2.5Sn	794	755
	Ti-3Al-2.5V	621	483
	Ti-8Al-1Mo-1V	897	828
	Ti-6Al-2Sn-4Zr-2Mo-0.1Si	931	862
α-β-Ti	Ti-6Al-4V	897	828
	Ti-4Al-4Mo-2Sn-0.5Si	1104	959
	Ti-6Al-6V-2Sn	1035	966
	Ti-6Al-2Sn-4Zr-6Mo	1172	1069
β-Ti	Ti-3Al-8V-6Cr-4Zr-4Mo	1172	1104
	Ti-15Mo-3Al-2.7Nb-0.2Si	792	750
	Ti-10V-2Fe-3Al	1241	1104
	Ti-15V-3Cr-3Sn-3Al	1000	966

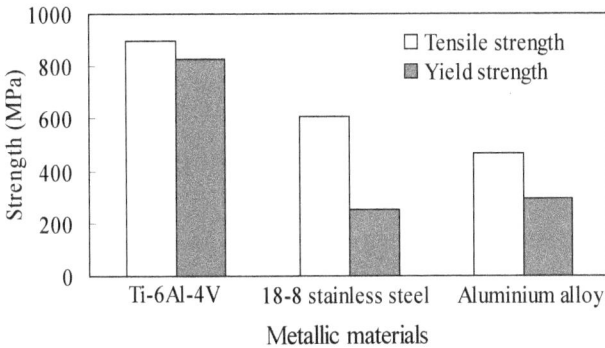

Figure 1-6 Tensile and yield strength of different metallic materials

Table 1-5 Elastic modulus of titanium alloys

Type of titanium		Elastic modulus (GPa)
CP-Ti	ASTM Grade 1	103
	ASTM Grade 2	103
	ASTM Grade 3	103
	ASTM Grade 4	104
α- and near α-Ti	Ti-3Al-2.5V	124
	Ti-8Al-1Mo-1V	198
	Ti-6Al-2Sn-4Zr-2Mo-0.1Si	191
α-β-Ti	Ti-6Al-4V	170
	Ti-4Al-4Mo-2Sn-0.5Si	160
β-Ti	Ti-15Mo-3Al-2.7Nb-0.2Si	135
	Ti-15V-3Cr-3Sn-3Al	147

3) Hardness

Hardness can be used to give a rough indication of the quality of titanium. The greater the hardness is, the higher the content of impurities, which results in worse quality. The effect of different impurities on the hardness of titanium is not the same. The largest influence is from nitrogen, oxygen and carbon, then iron, cobalt, and silicon, et al. The hardness of commercially pure titanium is about HV 150-200 MPa, titanium alloys are usually not more than HV 350 MPa.

1.3.6 Chemical Properties

The chemical activity of titanium at room temperature is very low, and it only reacts with several types of substance, such as fluoride, hydrofluoric acid, etc. However, at higher temperatures, it can react with a variety of elementary substances and compounds.

1.3.6.1 Chemical Reaction of Titanium with Elementary Substances

The elements can be divided into four classes according to their reaction with titanium (Figure 1-7).

Elements that react with titanium to form intermetallic compounds and limit solid solution
Elements that react with titanium to form all proportional solid solution
Elements that react with titanium to form covalent and ionic compounds
Elements that do not react with titanium

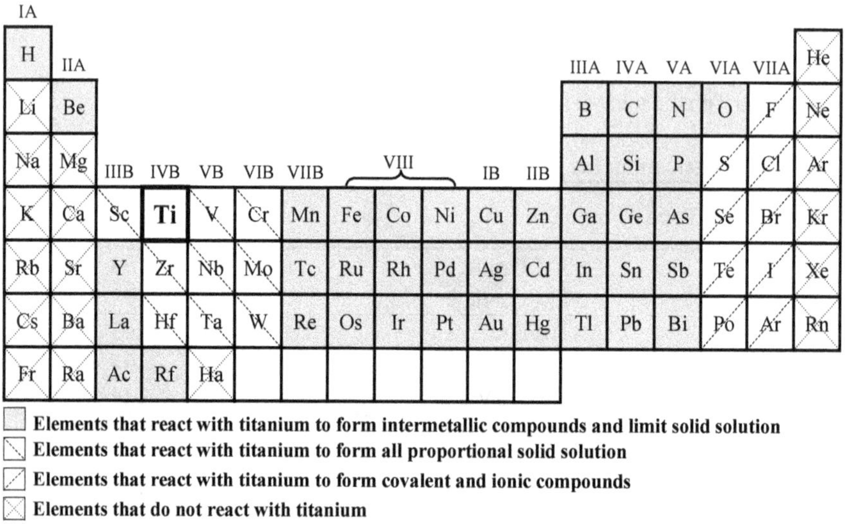

Figure 1-7 Classification of elements in periodic table based on the reaction with titanium (After Wang and Tian 2007)

1) Class I elements include the transition elements, hydrogen (H), beryllium (Be), boron (B) group, carbon (C) group and nitrogen (N) group elements. They react with titanium to form intermetallic compound and limit solid solution;

2) Class II elements include zirconium (Zr), hafnium (Hf), scandium (Sc), vanadium (V) group and chromium (Cr) group elements. They combine with titanium to form all proportional solid solution;

3) Class III elements include halogen and oxygen (O) group elements. They react with titanium to form covalent and ionic compound; and

4) Class IV elements include inert gas, alkali metal, alkaline earth metal, rare earth elements (except scandium Sc, actinium Ac and thorium Th). They normally do not react with titanium.

Due to the reaction characteristic of the metal elements in Class I and II with titanium, some of them are added into titanium as a strengthener or a stabilizer to control the microstructure and properties of titanium alloys. The detailed information can be found in the current published books on titanium alloys.

The reactions of the non-metal elementary substances in Class I and III with titanium are briefly summarized as follows.

When heated, titanium has high reactivity toward halogen, oxygen, nitrogen, hydrogen, and sulfur (Huang et al 2009). The active surface of titanium at room temperature can absorb hydrogen, and at 300°C, the amount of hydrogen absorption is very high. The temperature at which titanium starts to react obviously with oxygen is 600°C, and it is higher than 700°C for obvious reaction of titanium with nitrogen. Titanium can explode. Self-combustion temperature of dry titanium powder is 300-600°C. Titanium with the form of powder, sponge, dust and fines is easy to burn in the presence of sparks or a small flame. Titanium powder and filaments can be burned in nitrogen gas and chlorine gas.

1) Halogen

Titanium can react with all halogen elements, generating a titanium halide.

At ambient temperature, titanium can react with fluorine to form TiF_4. Intense chemical reaction takes place at 150°C.

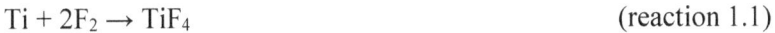

$$Ti + 2F_2 \rightarrow TiF_4 \hspace{4cm} \text{(reaction 1.1)}$$

At ambient temperature, titanium can also react with chloride. Intense chemical reaction occurs above 300°C.

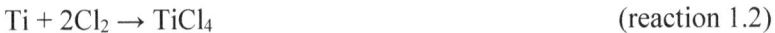

$$Ti + 2Cl_2 \rightarrow TiCl_4 \hspace{4cm} \text{(reaction 1.2)}$$

At 250-360°C, titanium can react with bromine.

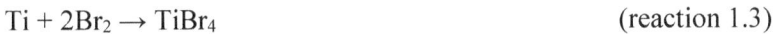

$$Ti + 2Br_2 \rightarrow TiBr_4 \hspace{4cm} \text{(reaction 1.3)}$$

At 170°C titanium can react with iodine to form TiI_4 gas and at 400°C the reaction accelerates.

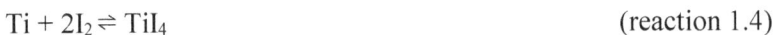

$$Ti + 2I_2 \rightleftharpoons TiI_4 \hspace{4cm} \text{(reaction 1.4)}$$

When the temperature exceeds 1000°C, TiI_4 can be decomposed into titanium and iodine, so it is a reversible reaction.

Halogen containing water has smaller effect on titanium than dry halogen, for example, wet chlorine with saturated water does not react with titanium below 80°C.

2) Oxygen

The reaction between titanium and oxygen depends on the existing pattern of titanium and temperature. Violent combustion or explosion takes place for powder titanium in air at room temperature in the presence of static electricity, sparks and friction. However, dense titanium is very stable in air at room temperature. When it is heated in air, it begins to react with oxygen, which diffuses into titanium lattice, and then forms a layer of compact oxide film. This oxide layer on the surface can prevent oxygen from penetrating into the titanium substrate and has a protective effect. Accordingly, titanium is stable in air below 500°C. Alloy elements, such as aluminum, tungsten and tin can reduce the oxidation rate of titanium, whereas zirconium raises the oxidation rate.

The reaction of titanium with oxygen in air is very slow at temperatures less than 100°C, and only the surface is oxidized at 500°C. With the increase of temperature, the surface oxide film is dissolved in titanium and oxygen diffuses towards the inner lattices of titanium. At temperatures of more than 700°C, oxygen diffusion accelerates, and the surface oxide film loses its protective effect. If the temperature reaches 1200-1300°C, titanium reacts violently with oxygen in air.

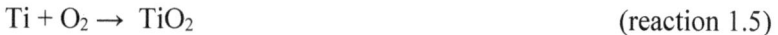

$$Ti + O_2 \rightarrow TiO_2 \qquad\qquad\qquad (reaction\ 1.5)$$

In pure oxygen, the occurring temperature of the intense reaction between titanium and oxygen is lower than that in air, approximately 500-600°C.

When the content of oxygen in titanium is more than the solubility limit, various kinds of titanium oxide, such as Ti_3O, Ti_2O_3, TiO, Ti_3O_5, TiO_2, etc, are generated. Oxygen enters the crystal lattice of titanium in the form of oxide, resulting in the significant increase in phase transition temperature (α-Ti$\rightarrow\beta$-Ti). Therefore, oxygen is a stabilizer of α-Ti. The maximum solubility of oxygen in the α-Ti (mass fraction) is 14.5%, and 1.8% in the β-Ti.

3) Nitrogen

Titanium does not react with nitrogen at room temperature. But at elevated temperatures, titanium is one of the few metals that can be burned in nitrogen gas, and the combustion temperature is about 800°C. The reaction between molten titanium and nitrogen is very intense. The reaction products include nitride (Ti_3N, TiN, etc) and Ti-N solid solution.

At 500-550°C, titanium absorbs nitrogen significantly, forming a solid solution. The nitrogen absorption rate increases when the temperature is higher than 600°C. Nitrogen with the form of titanium nitride (e.g. Ti_3N) goes into the crystal lattice of titanium in Ti-N solid solution, causing a rise of phase transition temperature (α-Ti$\rightarrow\beta$-Ti), thus nitrogen is also a stabilizer of α-Ti. The maximum solubility (mass fraction) of nitrogen in the α-Ti at 1050°C is 7%, and 2% in β-Ti at 2020°C. Nevertheless, the rate of nitrogen absorption is much slower than that of oxygen absorption, so titanium mainly absorbs oxygen in air, while nitrogen absorption is secondary.

4) Hydrogen

Ti-H solid solution and TiH, TiH_2 compounds are produced from reacting titanium with hydrogen. Hydrogen is easily soluble in titanium and 1 mol titanium can absorb almost 2 mol hydrogen. The hydrogen absorption rate and amount are related to temperature and hydrogen pressure. The hydrogen absorption is less than 0.002% at ambient temperature. There is a rise in the hydrogen absorption rate at 300°C. The maximum value appears at 500-600°C. Subsequently, with the increase of temperature, the amount of hydrogen absorption declines. At 1000°C, most hydrogen absorbed by titanium is resolved. A rise in hydrogen pressure can increase hydrogen absorption. Vice versa, a fall in hydrogen pressure can lead to dehydrogenation. Hence the reaction between titanium and hydrogen is reversible.

When titanium reacts with hydrogen, the small size hydrogen atoms quickly diffuse into the crystal lattice of titanium to generate an interstitial solid solution. The dissolution of hydrogen in titanium can reduce the phase transition temperature (α-Ti$\rightarrow\beta$-Ti), and the hydrogen is a stabilizer of β-Ti. When oxide film exists on the titanium surface, the rate of hydrogen absorption and dehydrogenation are greatly decreased.

5) Sulphur and phosphorus

At room temperature, sulphur does not react with titanium. At elevated temperatures, titanium sulfides, such as Ti_3S, Ti_2S, TiS, Ti_3S_4, Ti_2S_3, Ti_3S_5, TiS_2 and TiS_3, are produced by reaction of titanium with molten sulphur or gaseous sulphur.

Titanium and gaseous phosphorus can react when the temperature is higher than 450°C, forming Ti_2P below 800°C and TiP above 850°C.

6) Carbon and silicon

Titanium reacts with carbon only at elevated temperatures. The products comprise TiC and Ti-C solid solution. The presence of carbon in titanium can also increase the phase transition temperature (α-Ti$\rightarrow\beta$-Ti). The solubility of carbon in titanium is small, decreasing sharply with a drop of temperature. The maximum solubility (mass fraction) is 0.48% at 900°C. The solubility of carbon in β-Ti (mass fraction) has the highest value 0.8% at 1750°C. There is a limited solubility of carbon in α-Ti and β-Ti, so a large carbon content in titanium would cause the formation of dissociative titanium carbide phase in the microstructure.

As for silicon, titanium can react with it at high temperatures to produce silicon compounds having a high melting point, such as $TiSi_2$, $TiSi$ and Ti_5Si_3.

1.3.6.2 Chemical Reaction of Titanium with Compounds

Titanium oxidizes immediately upon exposure to air since it has a strong affinity for oxygen. This oxide film spontaneously formed on the titanium surface is a dense, adherent and stable passive film that protects the substrate from corrosion. This film is easily regenerated when subjected to mechanical wear or corrosion damage, indicating that titanium has a strong tendency to passivate. Therefore, the reactions of titanium with compounds are largely govered by the integrity of this oxide film (Boyer 2010).

Normally, in corrosive media containing oxidizing, neutral or weakly reducing compounds, such as sea water, chlorides, oxidizing inorganic acids, organic acids, hydrochloric acid with concentrations lower than 7%, and sulfuric acid with concentrations lower than 5%, the oxide film is stable and titanium is capable of withstanding corrosion attack (Wang and Tian 2007). Titanium is not corrosion resistant in media containing reducing acids or strong oxidizing substances. Four kinds of inorganic acid (hydrofluoric acid, hydrochloric acid, sulfuric acid and phosphoric acid), four kinds of hot concentrated organic acid (oxalic acid, formic acid, trichloroacetic acid and trifluoro acetic acid), and strongly corrosive aluminum chloride have severe corrosion effects on titanium. In strong oxidizing media, original oxide film may be destroyed, causing intense oxidation reaction on the titanium surface. This is why titanium can suddenly explode after a long period of contact with fuming nitric acid.

The media where titanium has high resistance to corrosion at room temperature are shown in Table 1-6.

Table 1-6 Media where titanium has high resistance to corrosion (After International Titanium Association 2011)

Category of Medium	Compounds
Water	Natural, sea, bracksih and polluted water
Inorganic oxidizing acids	Nitric, chromic, perchloric and hypochlorous acids
Inorganic salts	Chlorides of sodium, potassium, magnesium, calcium, copper, iron, ammonia, manganese and nickel
Organic acids	Acetic, terephthalic, adipic, citric, formic, lactic, stearic, tartaric and tannic acids
Organic compounds	Alchhols, aldehydes, esters, ketones and hydrocarbons, with air or moisture
Alkalis	Hydroxides of sodium, potassium, calcium, magnesium and ammonia
Gases	Sulfur dioxide, ammonium, carbon dioxide, hydrogen sulfide

More details about the corrosion behavior and corroiosn data of titanium alloys in different media are described in Chapter 2.

1.4 Present Situation of Titanium Resources

1.4.1 Titanium Reserves

Titanium is extracted from a number of titanium-bearing ores that occur naturally in the Earth, including rutile, ilmenite, leucosphenite, anatase, brookite, perovskite, titanite, etc (Metalpedia, 2017). *Rutile* has about 95% titanium dioxide. This is an important mineral raw material to refine titanium, but has less reserve in the crust. *Leucosphenite* has a titanium dioxide content of 70% to 92%. *Ilmenite* and *perovskite* contain 35%-52% titanium dioxide, but have large reserve in the crust and are also important raw ore resources for the production of metal titanium and titanium pigment. Table 1-7 lists the main titanium-containing mineral sources.

At present titanium is primarily produced from ilmenite (accounting for 90% of production) and rutile. Ilmenite deposits exist in China, Australia, Canada, India, South Africa, Brazil, Mozambique, Norway, Ukraine and United States, while rutile deposits are found in Australia, South Africa, India and Sierra Leone (Metalpedia 2017).

Table 1-7 Main mineral sources

Ore	Chemical formula	Content of TiO_2 (mass. %)	Density (g/cm^3)
Ilmenite	$FeTiO_3$	52.66	4.5-5.6
Rutile	TiO_2	100	4.5-5.2
Anatase	TiO_2	100	3.8-3.9
Brookite	TiO_2	100	3.7-4
Leucosphenite	$TiO_2 \cdot nH_2O$	about 94	3.5-4.5
Perovskite	$CaTiO_3$	58.9	3.9-4
Titanite	$CaTiSiO_3$	40.8	3.4-3.6
Pseudobrookite	Fe_2TiO_3	33.35	—
Arizonite	$Fe_2O_3 \cdot 3TiO_2$	60.01	4.25

According to USGS (U.S. Geological Survey) in 2014 (Figure 1-8 and 1-9), the total world reserves of ilmenite and rutile are about 700 million tonnes and 48 million tonnes, respectively. China, with 200 million tonnes of ilmenite, making up about 29% of the world total, is now the country that is most abundant in terms of ilmenite reserves. Australia has the largest rutile reserve in the world, accounting for half of the world total.

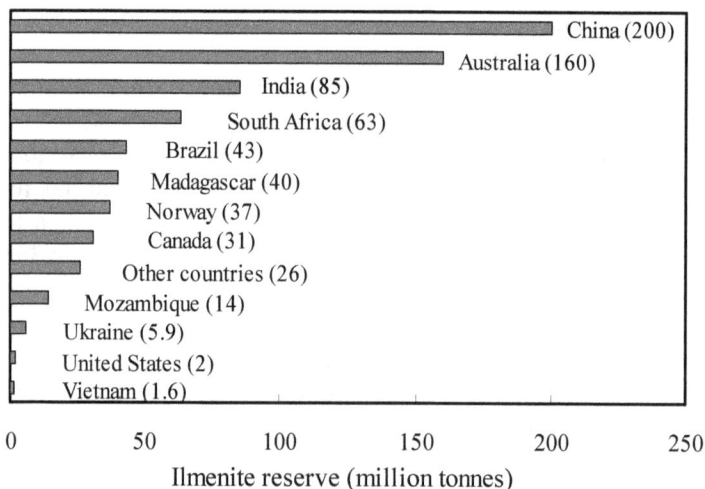

Figure 1-8 World ilmenite mineral reserves in 2014 (Data in million metric tons of contained TiO_2)

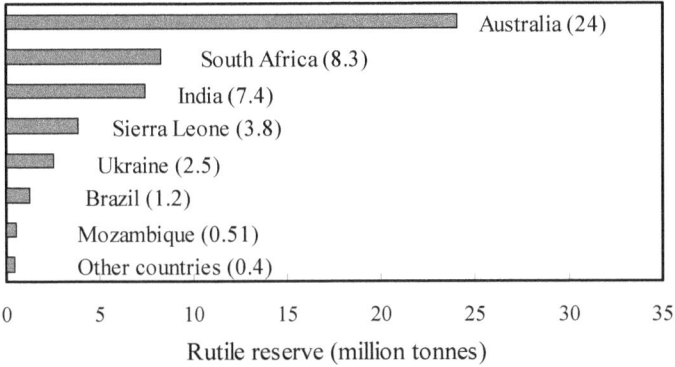

Figure 1-9 World rutile mineral reserves in 2014 (Data in million metric tons of contained TiO_2)

Based on the data of USGS, in 2012 and 2013, the total world ilmenite production is around 6,500 and 6,790 thousand tonnes, respectively (Figure 1-10), while rutile production is less, approximately 730 and 770 thousand tonnes, respectively (Figure 1-11).

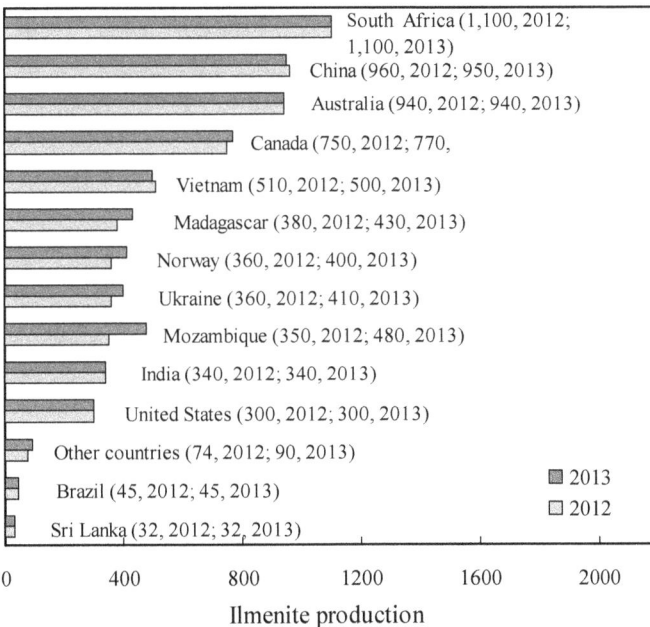

Figure 1-10 World ilmenite production in 2012 and 2013 (Data in thousand metric tons of contained TiO_2)

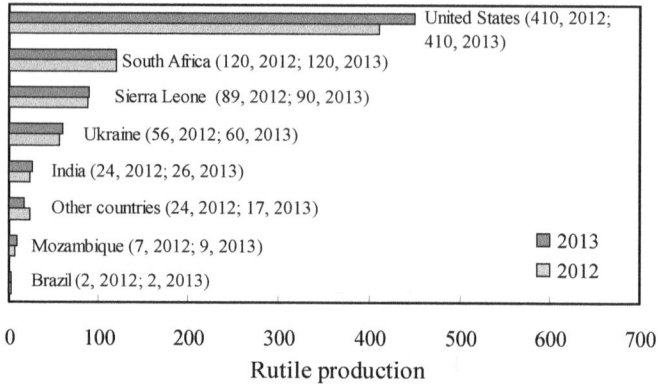

Figure 1-11 World rutile production in 2012 and 2013 (Data in thousand metric tons of contained TiO_2)

1.4.2 Titanium Applications

Titanium alloys are known for their high strength, low density, and excellent corrosion resistance, which make them attractive for a wide variety of industrial and commercial fields (Lin and Du 2014, AZO Material 2002).

Figure 1-12 depicts titanium usage in the 20th century in North America and Western Europe. The usage of titanium in the field of commercial aviation is the largest, accounting for 33%, followed by

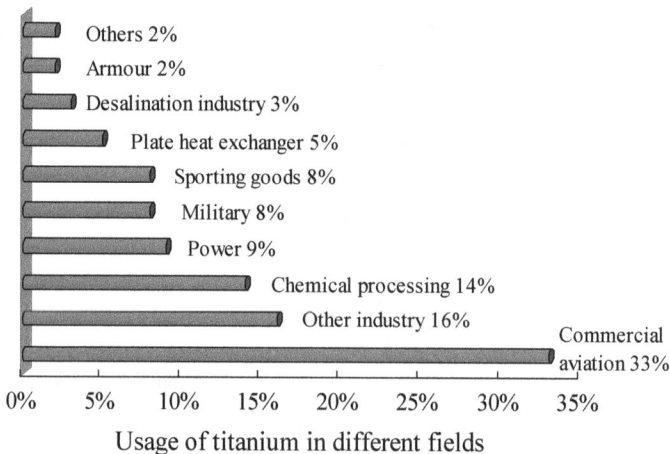

Figure 1-12 Usage of titanium in the 20th century in North America and Western Europe

chemical processing, power, military and sporting goods. A small portion is applied in plate heat exchanger, desalination industry and armour, etc.

Comparison of applications of titanium products in the United States and Japan is shown in Figure 1-13. The total usage amount of titanium products in the United States in 2001 is 24,979 tons (t), among which the application in aerospace is 14,500 t (58.5% of total usage), suggesting that titanium products are mainly used in aerospace industry. In Japan, the situation is different. Aerospace is not the major application field. The most common use of titanium products in Japan is in chemical and civil industries.

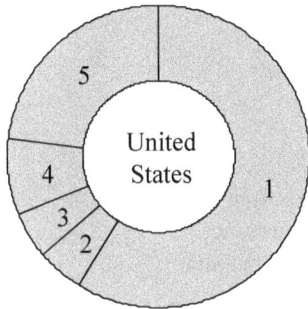

1-	Aerospace	58.5%
2-	Civil goods	5%
3-	Medical industry	5%
4-	Chemical industry, heat and nuclear power, sea water desalinization	8%
5-	Plate heat exchanger, automobile, vessel and marine, energy, architecture, sporting equipment, merchandising and others	23%

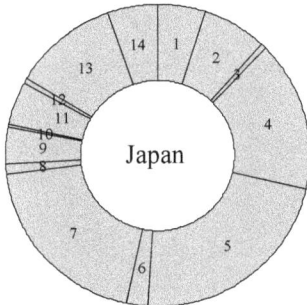

1-	Aerospace	4.9%
2-	Civil goods	7.0%
3-	Medical industry	5%
4-	Chemical industry	16.2%
5-	Geothermal and nuclear power	22.3%
6-	Sea water desalinization	2.4%
7-	Plate heat exchanger	19.6%
8-	Vessel and marine	1.1%
9-	Automobile	4.0%
10-	Energy	0.3%
11-	Sporting equipment	4.7%
12-	Architecture	0.6%
13-	Merchandising	11.0%
14-	Others	5.3%

Figure 1-13 Comparison of applications of titanium products in the United States and Japan

In China, titanium products are used primarily in the chemical industry. The usage percentage of titanium in various applications from 2003 to 2005 is depicted in Figure 1-14.

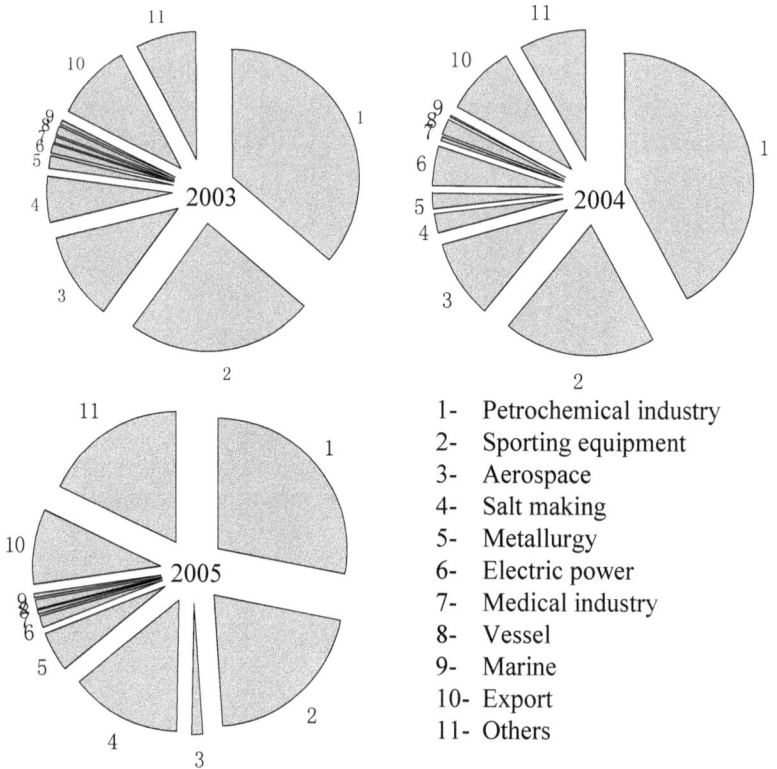

1- Petrochemical industry
2- Sporting equipment
3- Aerospace
4- Salt making
5- Metallurgy
6- Electric power
7- Medical industry
8- Vessel
9- Marine
10- Export
11- Others

Figure 1-14 Applications of titanium in China

Chapter 2 – Corrosion

2.1 Introduction

Titanium, with standard electrode potential, -1.63 V, is highly reactive and thermodynamically tends to be corroded in a medium. But as mentioned in Chapter 1, actually, titanium in a broad range of environments is very stable due to the fact that it readily develops a compact, inert oxide film with strong adhesion on the surface. Hence, titanium substrate is protected. Even after the film is broken down, it repairs itself instantly, so titanium is generally self-passivating. The outstanding performance of titanium alloys is excellent for corrosion resistance in fresh and sea water. It exhibits better resistance to corrosion than aluminum alloys, stainless steel and nickel based alloys in sea water. At room temperature titanium dose not react with chlorine, dilute sulphuric acid, dilute hydrochloric acid, nitric acid and chromic acid; and shows good corrosion resistance in aqueous alkalis, many organic acids and chemical compounds; but can be corroded by hydrofluoric acid, borofluoric acid, fluorosilicic acid, hydrochloric acid with concentrations higher than 7%, sulfuric acid with concentrations higher than 5%, phosphoric acid and molten alkalis.

In this chapter, anticorrosion performance and some corrosion data of titanium alloys in various service environments, mechanism and influencing factors for different forms of corrosion are presented. In addition, the suggestions for material selection, use design and prevention measures are proposed (Lin and Du 2014).

2.2 Anticorrosion Performance in Service Environments

2.2.1 Atmospheric Environment

Titanium and its alloys exhibit excellent corrosion resistance in atmosphere as a result of the protective oxide film on the surface. After 24 years of exposure in marine atmosphere, the average corrosion rate of titanium was smaller than 0.254 μm/y (micron per year) (Huang and Zuo 2003), which is almost negligible, and similar corrosion data were observed in industrial and agricultural atmospheres.

2.2.2 Marine Environment

Due to the presence of various ions, such as Na^+, K^+, Mg^{2+}, Cl^-, etc., seawater is very corrosive. Generally, materials used in seawater suffer from corrosion due to the marine atmosphere, seawater, tide (alternate effect of seawater and atmosphere) and marine organism, as well as seawater erosion and periodic wave impact. Accordingly materials used in such environment should have good corrosion resistance and comprehensive mechanical properties. Moreover, low density and high strength are necessary for materials to be used for manufacturing floating equipment. Steel, stainless steel, copper and aluminum are not corrosion resistant materials in seawater, and must be coated or protected for marine utilization. Titanium is the only structural material that has outstanding corrosion resistance without using a protective coating in marine environment.

As stated earlier, the standard electrode potential of titanium is -1.63 V, but passivation causes a sharp positive shift in corrosion potential. At 25°C the corrosion potential of titanium in seawater is about 0.35 V (vs. SHE). The passive film on the surface of titanium is highly self-healable, and can be repaired quickly with a newly formed film after breakdown. This contributes to the unique corrosion resistance of titanium in marine environment.

Zhu (1995) indicated no obvious corrosion was observed on the surface after years of exposure at different depths of seawater. The corrosion rates of CP-Ti and Ti-6Al-4V in both shallow and deep sea are given in Table 2-1. The unit of corrosion rate is mm/y (milimeter per year).

Table 2-1 Corrosion rate of CP-Ti and Ti-6Al-4V in shallow and deep sea (After Zhu 1995)

Type of titanium	Depth of water (m)	Corrosion rate (mm/y)
CP-Ti	shallow sea	8×10^{-7}
	1720	4.0×10^{-5}
Ti-6Al-4V	1720	8.0×10^{-6}

As shown in Table 2-2, Ti-6Al-4V has good corrosion resistance in flowing seawater. When the speed of flowing seawater is smaller than 8 m/s, the corrosion rate is negligible. Even if the speed is increased to 36 m/s, it is lower than 0.01 mm/y.

Table 2-2 Corrosion rate of Ti-6Al-4V in still and flowing seawater (After Huang et al 2009)

Seawater	Corrosion rate (mm/y)
Still seawater	7×10^{-5}
Flowing seawater (<8 m/s)	0

Long term corrosion tests on Ti-55, Ti-4Al-0.005B and Ti-6Al-4V in seawater were carried out in China (Huang et al 2009), and the results also showed that they have extraordinary corrosion resistance in flowing seawater.

1)   No corrosion was detected on Ti-4Al-0.005B in seawater with flowing speed which was less than 7.45 m/s.

2)   Very slight corrosion, about 0.0003 mm/y, occurred on Ti-55 and Ti-4Al-0.005B in seawater flowing with a velocity of 10 m/s.

3)   Slight corrosion, less than 0.001 mm/y, was observed on these titanium materials in flowing seawater at 36 m/s velocity.

Titanium has good resistance to general and local corrosion in marine environment. It does not corrode in splash and tide zone, as well as under scouring of high-speed seawater. Sediment on the titanium surface does not cause crevice corrosion and pitting corrosion, and the corrosion resistance is not affected by sulfide in seawater. It is estimated (Huang et al 2009) that corrosion depth of titanium in sulfide contaminated still seawater is less than 0.03 mm over a thousand years, which is almost negligible. Suspended abrasive particles in seawater may cause damage to some metals, such as copper or aluminum alloys, but not to titanium. In seawater, it is very common that marine organisms adhere to titanium because it is not toxic to them. However, the marine organisms do not cause crevice corrosion and pitting corrosion where they cover, because the corrosion resistant oxide film beneath is intact.

Titanium is the preferred material to use for ocean engineering, regardless of whether the seawater is still or moving with high velocity, clean or polluted, the passive film is stable even in seawater containing plenty of mud and sand.

Titanium is corrosion resistant in neutral seawater below 130°C. Only in seawater with high temperatures or low pH pitting corrosion or localized corrosion may occur.

2.2.3 Industrial Environments

As mentioned in section 1.3.6.2, generally, it is corrosion resistant in a medium where passive film on the surface is stable, such as oxidizing, neutral, or weakly reducing media. However, it corrodes in a media that can destroy the passive film, such as in a reducing acid solutions.

2.2.3.1 Inorganic Acids

Titanium has good corrosion resistance in oxidizing acid solutions, for example, nitric acid and chromic acid. In reducing acid solutions, such as diluted hydrochloric acid, sulfuric acid or phosphoric acid, the dissolution of titanium is much slower than iron, but it accelerates with the rise in concentration and temperature. In hydrochloric acid mixed with nitric acid, titanium dissolves quickly.

1) Nitric acid

Nitric acid is an oxidizing acid. In this acid, a compact oxide film is easily formed and maintained on titanium surface. As the concentration of nitric acid increases gradually, the interference color of the film varies from pale yellow, light yellow, earth yellow, brownish yellow, finally to blue. This film provides a formidable corrosion-resistant barrier, and accordingly the corrosion rate of titanium alloy in nitric acid is very low. So, it is often used in nitric acid systems. But titanium with rough surfaces, particularly sponge titanium and powder titanium, can react with cold or hot nitric acid (Huang et al 2009).

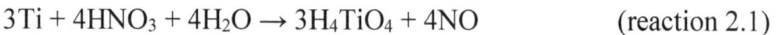

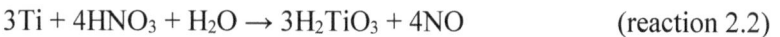

$$3Ti + 4HNO_3 + 4H_2O \rightarrow 3H_4TiO_4 + 4NO \qquad \text{(reaction 2.1)}$$

$$3Ti + 4HNO_3 + H_2O \rightarrow 3H_2TiO_3 + 4NO \qquad \text{(reaction 2.2)}$$

The effect of different temperatures and acid concentrations on the corrosion of CP-Ti is shown in Table 2-3. Temperature increments promote the corrosion of titanium in nitric acid. When the temperature is higher than 70°C, titanium reacts with nitric acid, and the corrosion rate reaches 9-9.5 mm/y in 30%-35% nitric acid at 230°C. The concentration rise of nitric acid also enhances the corrosion rate. However, when the concentration reaches a certain value, nitric acid facilitates the passivation of titanium instead and strengthens the corrosion resistance.

Table 2-4 gives the corrosion rate of CP-Ti in nitric acid without and with aeration of air and in boiling nitric acid.

Table 2-3 Corrosion rate of CP-Ti in 20%-60% nitric acid at elevated temperatures (unit: mm/y) (After Zhang et al 2005)

Temperature	Concentration of nitric acid					
(°C)	20%	30%	35%	40%	50%	60%
190	1.4	2.3	2.6	2.5	1.8	—
200	3.5	4.9	5.2	5.1	4.2	2.5
210	4.6	6.1	6.5	6.3	5.2	2.0
220	5.8	7.7	8.2	8.1	6.5	3.0
230	6.7	9	9.5	9.2	7.3	4.0

Table 2-4 Corrosion rate of CP-Ti in nitric acid with and without aeration of air and in boiling nitric acid (After Wang and Tian 2007)

Status	Concentration (%)	Temperature (°C)	Corrosion rate (mm/y)
	10	room temperature	0.005
	20	room temperature	0.245
	30	room temperature	0.004
	40	room temperature	0.002
	50	room temperature	0.002
	60	room temperature	0.001
	70	room temperature	0.005
Aerated	10	40	0.003
	20	40	0.005
	30	50	0.015
	40	50	0.016
	50	60	0.037
	60	60	0.040
	70	70	0.040
	40	200	0.610
	70	270	1.220
Not aerated	35	80	0.051-0.102
	70	80	0.025-0.076
	17	boiling	0.076-0.102
—	35	boiling	0.127-0.508
	70	boiling	0.640-0.940

Ions of titanium and some heavy metals are corrosion inhibitors to titanium in acid. Figure 2-1 shows the effect of titanium ion on the corrosion rate of CP-Ti in boiling nitric acid.

A small amount of silicon compound can retard the corrosion of titanium in nitric acid at elevated temperatures. For example, polysiloxane oil was added to 40% nitric acid at an elevated temperature and it almost reduced the corrosion rate to zero (Zhang et al 2005). However, phosphide can accelerate corrosion in nitric acid, and it can be used to prepare acid pickling solution of titanium.

Some other compounds added into nitric acid also have an impact on the corrosion rate of titanium (Table 2-5).

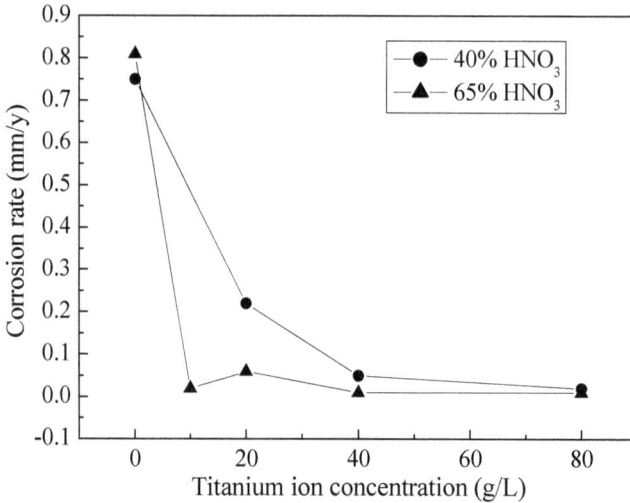

Figure 2-1 Effect of titanium ion concentration on corrosion rate of CP-Ti in boiling 40% and 65% HNO_3 (After Zhu 1995)

Table 2-5 Corrosion rate of CP-Ti in nitric acid with addition of other compounds (After Wang and Tian 2007)

Solution	Concentration (%)	Temperature (°C)	Corrosion rate (mm/y)
+0.1%CrO_3	40	boiling	0.003-0.025
+10%$FeCl_3$	40	boiling	0.122-0.188
+0.1%$K_2Cr_2O_7$	40	boiling	<0.016
+10%$NaClO_3$	40	boiling	0.003-0.0360
+Saturated $ZrO(NO_3)_2$	33-45	118	0
+15%$ZrO(NO_3)_2$	65	127	0
+179g/L $NaNO_3$ + 32g/L NaCl	20.8	boiling	0.127-0.295
+170g/L $NaNO_3$ +2.9g/L NaCl	27.4	boiling	0.483-2.920

2) Hydrochloric acid

Hydrochloric acid is classified as a reducing acid, in which titanium is not stable even at room temperature. The corrosion rate of titanium increases with an increase in hydrochloric acid concentration and temperature (Mogoda and Ahmad 2004). Titanium alloy is commonly used in 3% hydrochloric acid at room temperature, or 0.5% hydrochloric acid at 100°C. Table 2-6 lists the corrosion rate of CP-Ti, Ti-6Al-4V, Ti-0.8Ni-0.3Mo and Ti-0.2Pd in some diluted hydrochloric acid solutions. Ti-Ni-Mo and Ti-Pd have higher corrosion resistance than CP-Ti. At room temperature, CP-Ti, Ti-Ni-Mo and Ti-Pd can be used in hydrochloric acid with concentrations less than 7%, 9% and 27%, respectively.

Titanium is not corrosion resistant in hydrochloric acid, but this can be improved by alloying, anodic passivation and using corrosion inhibitor (Zhang et al 2005). There are two sorts of corrosion inhibitors for titanium in hydrochloric acid. One is a strong oxidizing inorganic inhibitor, including nitric acid, potassium dichromate, sodium hypochlorite, chlorine, oxygen and multivalent heavy metal ions (Fe^{2+}, Fe^{3+}, Cu^{2+}, some noble metal ions, etc). The other is an organic inhibitor, including oxidizing organic compounds, azoic compounds, quinine and anthraquinone compounds, heterocyclic compounds, complexing inhibitors.

Table 2-6 Corrosion rate of titanium alloys in dilute hydrochloric acid (from Zhu 1995)

Temperature (°C)	HCl (wt.%)	Corrosion rate (mm/y)			
		CP-Ti	Ti-6Al-4V	Ti-Ni-Mo	Ti-Pd
25	1	0	—	0.005	0.003
	2	0	—	0.003	0.006
	3	0.013	—	0.013	0.010
	5	0.005	—	0.013	0.015
	8	0.005	—	0.005	0.025
Boiling	1	2.16	—	0.036	0.020
	2	7.11	6.60	0.254	0.046
	3	14.0	13.2	10.2	0.069
	5	21.3	26.2	38.1	0.254
	8	>50.8	48.3	76.2	0.610

Multivalent heavy metal ions have an inhibition effect on titanium corrosion in acid solutions. For example, the corrosion resistance of CP-Ti is improved by the addition of high concentration Fe^{3+} to hydrochloric acid, which is almost close to the corrosion resistance of Ti-Pd or Ti-Ni-Mo. Other heavy metal ions, such as Cu^{2+}, Ni^{2+}, Mo^{6+}, Ti^{4+}, enhance the corrosion resistance by passivation in hydrochloric acid (shown in Table 2-7). Various kinds of oxidizing agents, such as nitric acid, chlorine, sodium hypochlorite and chromic acid, are also good inhibitors for titanium.

Table 2-7 Effect of multivalent heavy metal ions on corrosion rate of CP-Ti in 5% boiling hydrochloric acid (unit: mm/y) (After Huang and Zuo 2003)

Concentration (mg/kg)	Heavy metal ion				
	Fe^{3+}	Cu^{2+}	Mo^{6+}	Cr^{6+}	V^{5+}
0	29.0	29.0	29.0	29.0	29.0
100	0.025	0.033	0	0	0.020
500	0.020	0.030	0	0	0.008

3) Sulfuric Acid

Sulfuric acid is also a reducing acid, and titanium is resistant to corrosion in dilute sulfuric acid at low temperatures, for example, 20% sulfuric acid at 0°C and 5% sulfuric acid at room temperature. The corrosion rate increases with an increase in temperature. If the temperature is raised to 100°C, it is corrosion resistant in sulfuric acid below 0.2%. Some corrosion resistant titanium alloys show a better resistance to corrosion in sulphuric acid. Ti-0.2Pd is stable in 47% sulfuric acid at room temperature and 7% boiling sulfuric acid. Ti-0.8Ni-0.3Mo is stable in 1% boiling sulfuric acid (Zhang et al 2005)).

In sulfuric acid with the concentration lower than 4%, a protective oxide film forms on the surface of titanium, preventing further corrosion. When the concentration is higher than 5%, an obvious reaction can be observed. At room temperature, the highest corrosion rate occurs in 40% sulfuric acid. Diffluent $[Ti(SO_4)_{2+x}]^{2x-}$ ions are produced, which resolve into TiO_2 and H_2SO_4 when the concentration is higher than 40%. That is why the corrosion rate slows down in 60% sulfuric acid.

Titanium may react with hot dilute sulfuric acid or 40% sulfuric acid, producing titanous sulfate (Huang et al 2009, Seiji and Yukari 2006).

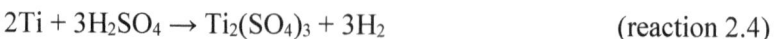

$$Ti + H_2SO_4 \rightarrow TiSO_4 + H_2 \qquad \text{(reaction 2.3)}$$

$$2Ti + 3H_2SO_4 \rightarrow Ti_2(SO_4)_3 + 3H_2 \qquad \text{(reaction 2.4)}$$

Hot concentrated sulfuric acid reacts with titanium and produces SO_2:

$$2Ti + 6H_2SO_4 \rightarrow Ti_2(SO_4)_3 + 3SO_2 + 6H_2O \qquad \text{(reaction 2.5)}$$

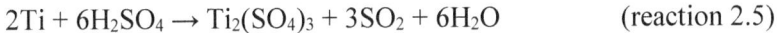

Generally, chlorine mitigates titanium corrosion in sulfuric acid, but at 90°C in 50% sulfuric acid it accelerates the corrosion, and even may cause ignition. Table 2-8 and Table 2-9 suggest that the corrosion resistance is improved by adding multivalent heavy metal ions, such as Fe^{3+}, Mo^{6+}, Cr^{6+}, Cu^{2+}, Ti^{4+}, and oxidizing agents. Some organic corrosion inhibitors, such as α-nitroso-β-naphthol, p-nitrobenzoic acid, paranitroaniline, picric acid, quinine, anthraquinone compound, are also effective, as well as some clathrate inhibitors. Normally, titanium is seldom used in sulfuric acid system.

Table 2-8 Effect of multivalent heavy metal ions on corrosion of CP-Ti in boiling 10% sulfuric acid (unit: mm/y) (After Huang and Zuo 2003)

Concentration (mg/kg)	Heavy metal ion				
	Fe^{3+}	Cu^{2+}	Mo^{6+}	Cr^{6+}	V^{5+}
0	>76.2	>76.2	>76.2	>76.2	>76.2
100	0.208	0.419	0.001	0.001	0.005
500	0.069	0.361	0.000	0.001	0.005

Table 2-9 Corrosion rate of CP-Ti in sulphuric acid (After Wang and Tian 2007)

Status	Concentration (%)	Temperature (°C)	Corrosion rate (mm/y)
	1	60	0.008
	3	60	0.013
	5	60	4.83
	1	100	0.005
	3	100	23.4
	5	100	20.6
Aerated with air	10	35	1.27
	40	35	8.64
	75	35	1.07
	75	boiling	154.5
	80	25	8.03
	80	boiling	189.5
	98	25	1.57
	98	boiling	5.38
Aerated with nitrogen	1	100	7.16
	3	100	21.1
	5	100	26.9

Table 2-9 Continued

Status	Concentration (%)	Temperature (°C)	Corrosion rate (mm/y)
—	1	boiling	17.8
—	5	boiling	25.4
+0.25%CuSO$_4$	5	93	0
	5	93	0
	30	38	0.061
	30	93	0.088
+0.5%CuSO$_4$	30	38	0.067
	30	93	0.823
+1.0%CuSO$_4$	30	38	0.020
	30	93	0.084
	30	boiling	1.650
+0.5%CrO$_3$	5	93	0
	30	93	0
+10%HNO$_3$	90	25	0.457
+30%HNO$_3$	70	25	0.635
+50%HNO$_3$	50	25	0.635
+70%HNO$_3$	30	25	0.102
+90%HNO$_3$	10	25	0
+90%HNO$_3$	10	60	0.011
+50%HNO$_3$	50	60	0.399
+20%HNO$_3$	80	60	1.590
+4.79g/L Ti^{4+}	40	100	0

4) Phosphoric acid

Phosphoric acid is a reducing acid, in which titanium has moderate corrosion resistance. Given the same temperature and acid concentration, the corrosion rate of titanium in phosphoric acid is lower than that of hydrochloric and sulfuric acid, but higher than nitric acid. It increases with temperature and acid concentration, similar to the case in hydrochloric acid.

CP-Ti is corrosion resistant in 30% phosphoric acid aerated with air at room temperature, 10% phosphoric acid at 60°C, and only 2% phosphoric acid at 100°C. Under the same circumstances Ti-0.2Pd is much more corrosion resistant than CP-Ti, and it is stable in 80% at room temperature, 15% at 60°C and 6% phosphoric acid at 100°C. The corrosion resistance of Ti-0.8Ni-0.3Mo is between CP-Ti and Ti-0.2Pd. Generally, titanium can be used in 30% phosphoric acid aerated with air at 20°C, or 20% phosphoric acid at 35°C without aeration (Zhu 1995).

Corrosion reaction between titanium and phosphoric acid occurs as follows (Zhang et al 2005):

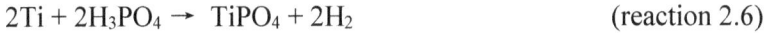

$$2Ti + 2H_3PO_4 \rightarrow TiPO_4 + 2H_2 \qquad \text{(reaction 2.6)}$$

Table 2-10 shows some corrosion data of CP-Ti in phosphoric acid. Oxidizing agents or other inhibitors improve corrosion resistance of titanium in phosphoric acid, such as Cu^{2+}, Fe^{3+}, HNO_3, chlorine, silver and mercury.

Table 2-10 Corrosion rate of CP-Ti in phosphoric acid (After Wang and Tian 2007)

Concentration (%)	Temperature (°C)	Corrosion rate (mm/y)
10-30	25	0.020-0.051
30-80	25	0.051-0.762
1	boiling	0.254
10	boiling	10.2
30	boiling	26.2
10	80	1.83
81+3%HNO$_3$+16%H$_2$O	88	0.381

5) Hydrofluoric acid

Hydrofluoric acid and fluosilicic acid are reducing acids. They are very strong aggressive corrosion media. Even very dilute hydrofluoric acid can cause severe corrosion on titanium at room temperature, as well as acid medium containing fluorine, such as fluosilicic and fluoboric acids. Hence, titanium cannot be applied in hydrofluoric acid system (Zhang et al 2005).

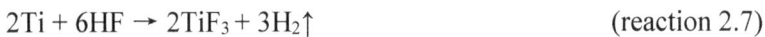

$$2Ti + 6HF \rightarrow 2TiF_3 + 3H_2\uparrow \qquad \text{(reaction 2.7)}$$

As a porous corrosion product, TiF_3 is not protective for the substrate and the corrosion quickly develops.

Hydrofluoric acid that is mixed with hydrochloric or sulfuric acid is more corrosive to titanium than itself. In these cases, complexation between F^- and Ti^{4+} accelerates titanium dissolution.

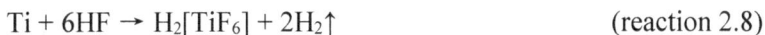

$$Ti + 6HF \rightarrow H_2[TiF_6] + 2H_2\uparrow \qquad \text{(reaction 2.8)}$$

When a small amount of soluble fluoride is added to other acids, such as hydrobromic acid, perchloric acid, formic acid or acetic acid, the corrosion rate of titanium increases several ten times. Acidic fluoride solutions, such as NaF, KHF_2, also cause serious corrosion on titanium. So far there is no ideal inhibitor for titanium in hydrofluoric acid.

The corrosion rate of titanium in hydrofluoric acid increases with temperature and acid concentration. If hydrofluoric acid is mixed with nitric acid, the concentration of nitric acid plays an important role in the corrosion rate of titanium, which at first increases with concentration, then decreases with concentration when the concentration reaches a threshold value (Lin et al 2008, Lin and Hong 2011).

Fe^{2+}, Ni^{2+}, Ag^+, Cu^{2+}, Au^{3+} and Pt^{4+} accelerate titanium corrosion in hydrofluoric acid. Mg^{2+} has no influence on the corrosion reaction, but Pb^{2+} slows down the corrosion. Very effective corrosion inhibitors for titanium in hydrofluoric acid have not been found yet.

2.2.3.2 Organic Acids

Corrosion of titanium in organic acid is dependent upon the relative magnitude of oxidability and reductibility of the medium. Table 2-11 gives the corrosion rate of CP-Ti in some organic acids.

Only a few organic acids can corrode titanium in deaerated condition, such as hot formic acid, hot concentrated oxalic acid, hot concentrated trichloroacetic acid and hot sulfamic acid. Aeration can improve corrosion resistance of titanium in most non-oxidizing acids. The corrosion rate in formic acid is obviously lowered after aeration. Titanium has good corrosion resistance in acetic acid, citric acid, tartaric acid, lactic acid and tannic acid and it can be used as equipment material in the chemical industry to produce adipic acid, terephthalic acid, and monochloro acetic acid, etc.

2.2.3.3 Organic Compounds

Titanium exhibits good corrosion resistance in organic compounds, such as alcohols, aldehydes, hydrocarbons. Hence it has been applied more and more widely in organic synthesis industry. Table 2-12 gives the corrosion rate of CP-Ti in some organic compounds.

As a structural material in organic synthesis production, it is more corrosion resistant than stainless steel in organic media.

In general, water and oxygen in organic media are in favor of passivation on titanium, and it is difficult to maintain its passivation in water-free organic media. For example, stress corrosion cracking of CP-Ti was observed when it was used in methyl alcohol with water content less than 1.5%. At elevated temperatures in water-free environments, possible hydrogen evolution resulting from organic compound decomposition may be noticed, because it may cause hydrogen absorption and hydrogen embrittlement of titanium.

2.2.3.4 Alkalis

In alkali media such as sodium hydroxide, potassium hydroxide, ammonium hydroxide and magnesium hydrate, titanium is in a passivation state and hence has good corrosion resistance. Table 2-13 and Table 2-14 list the corrosion rate of CP-Ti at different temperatures in NaOH and KOH, respectively. Almost no corrosion on titanium is observed in boiling saturated $Ca(OH)_2$, $Mg(OH)_2$ or NH_4OH.

The corrosion rate is low in alkalis, but in boiling alkali media (pH>12) titanium may absorb hydrogen.

Table 2-11 Corrosion rate of CP-Ti in some organic acids (After Huang and Zuo 2003)

Medium	Concentration (%)	Temperature (°C)	Corrosion rate (mm/y)
Acetic acid	5-99.5	100	0
Citric acid	50	100	<0.0003
Citric acid (aerated)	50	100	<0.127
Citric acid (deaerated)	50	boiling	0.356
Formic acid (aerated)	10-90	100	<0.127
Formic acid (deaerated)	10-90	boiling	>1.27
Lactic acid	10	60	0.003
	10	100	0.048
	85	100	0.008
Lactic acid (deaerated)	10	boiling	0.014
	25	boiling	0.028
	85	boiling	0.010
Oxalic acid	1	35	0.151
	1	60	4.50
	25	100	49.4
Octadecanoic acid	100	182	<0.127
Sulfamic acid (deaerated)	10	boiling	13.7
Tartaric acid	50	100	0.005
Tannic acid	25	100	0
Trichloroacetic acid	100	boiling	14.55

Table 2-12 Corrosion rate of CP-Ti in organic compounds (After Huang and Zuo 2003)

Medium	Concentration (%)	Temperature (°C)	Corrosion rate (mm/y)
Acetic anhydride	99-99.5	20-boiling	<0.127
Adipic acid +15-20%glutaric acid +acetic acid	25	193-200	0
Adiponitrile	vapour	371	0.008
Adipic chloride +chlorobenzene	—	—	0.003
Alcohol	91	35	0
	95	boiling	0.013
aldehyde	100	150	0
Aniline+2%AlCl$_3$	98	316	20.4
Aniline hydrochloride	5-20	35-100	<0.001
Benzene	liquid	25	0
Benzene+KCl, NaCl	liquid	80	0.005
Carbon tetrachloride	99	boiling	<0.127
Chloroform	100	boiling	0
Chloroform+water	—	boiling	0.127
Chlorylene	100	boiling	<0.127
Cyclohexane +trace amount formic acid	—	150	0.003
Formaldehyde	37	boiling	<0.127
Tetrachloroethylene	100	boiling	<0.127
Tetrachloroethane	100	boiling	<0.127
Trichloroethene	99	boiling	<0.003

2.2.3.5 Salts

Titanium is corrosion resistant in most salt solutions, particularly in oxidizing salt solutions. At low temperatures titanium dose not react with water-free fluoride, but corrodes severely in molten fluoride at elevated temperatures. A small quantity of soluble fluoride in acid solutions, such as nitric acid, perchloric acid, phosphoric acid, hydrochloric acid, sulfuric acid, may greatly accelerate the corrosion of titanium. However, high concentration fluoride may slow down the corrosion of titanium in sulfuric acid. The corrosion rate of titanium in chloride solutions is very low, even at elevated temperatures. Table 2-15 shows the corrosion rate of CP-Ti in aerated chloride solutions.

Table 2-13 Corrosion rate of CP-Ti in sodium hydroxide (Sources: Zhu 1995, Wang and Tian 2007)

Concentration (%)	Temperature (°C)	Corrosion rate (mm/y)
5-10	21	0.001
10	boiling	0.021
28	25	0.003
40	66	0.038
	93	0.064
	121	0.127
50	38	0.002
	57	0.0127
	66	0.018
73	110	0.051
	129	0.178
50-73	188	>1.09

Table 2-14 Corrosion rate of CP-Ti in potassium hydroxide (Sources: Zhu 1995, Wang and Tian 2007)

Concentration (%)	Temperature (°C)	Corrosion rate (mm/y)
10	103	0.13
25	108	0.30
	boiling	0.305
50	29	0.010
	boiling	2.74
50-100	241-377	1.02-1.52

Table 2-15 Corrosion rate of CP-Ti in aerated chloride solutions (After Zhu 1995)

Medium	Concentration (%)	Temperature (°C)	Corrosion rate (mm/y)
AlCl₃	5-10	60	0.003
	10	100	0.002
	10	150	0.033
	20	149	16.0
	25	20	0.001
	25	121	6.55
BaCl₂	5-25	100	0
CuCl₂	1-20	100	<0.013
	40	boiling	0.005
	40	90	0.005

Table 2-15 Continued

Medium	Concentration (%)	Temperature (°C)	Corrosion rate (mm/y)
CuCl	50	149	0.003
CaCl$_2$	5	100	0.001
	10	100	0.008
	20	100	0.015
	55	104	0.001
	60	149	0
	62	177	0.406
	73	177	2.13
FeCl$_3$	1-20	21	0
	1-40	boiling	<0.013
	50	boiling	0.004
	50	150	<0.018
HgCl$_2$	1	100	0
	5	100	0.011
	10	100	0.001
	saturated	70-194	<0.127
KCl	saturated	60	0
	saturated	boiling	0
LiCl	50	100	0
MgCl$_2$	5	100	0.001
	20	100	0.010
	50	199	0.005
MnCl$_2$	5-20	100	0
NiCl$_2$	5	100	0.004
	20	150	0.003
	50	200	0
	75	200	0.610
	80	104	203.2
NaCl	saturated	boiling	0
	pH 1.5+23	boiling	0
	pH 1.2+23	boiling	0.071
NH$_4$Cl	various concentrations	20-100	<0.013
ZnCl$_2$	20	150	0
	50	200	0
	75	200	0.610
	80	100	203.2

In most chloride solutions the corrosion rate is low, except in ZnCl$_2$, AlCl$_3$ and CaCl$_2$ solutions with high concentrations at elevated temperatures. The main corrosion form of titanium in chloride solution is crevice corrosion. Accordingly, it is the severity of crevice corrosion, instead of general corrosion phenomenon, that determines the application possibility of titanium material in chloride solutions if various forms of crevice exist. Corrosion resistant titanium alloys, such as Ti-Pd and Ti-Ni-Mo are appropriate to be applied in chloride solutions, as shown in Table 2-16.

Table 2-16 Crevice corrosion test results of titanium alloys in boiling chloride solutions for 500h ("+"----presence of crevice corrosion, "–"----no crevice corrosion) (After Huang and Zuo 2003)

Medium	pH	CP-Ti	Ti-0.8Ni-0.3Mo	Ti-0.2Pd
$ZnCl_2$	3.0	+	–	–
10%$AlCl_3$	—	+	–	–
10%$MgCl_2$	4.2	+	–	–
10%NH_4Cl	4.1	+	–	–
NaCl	1.0	+	–	–
10%Na_2CO_3	1.0	+	–	–
10%$FeCl_3$	0.6	+	+	–

2.2.3.6 Gaseous Environments

Dry chlorine and pure oxygen can induce combustion and explosion of titanium. However, if chroline contains water, strong oxidation can enable titanium to maintain passivation. In wet chlorine, the corrosion resistance of titanium is better than other commonly used metals. Titanium can absorb hydrogen from the environment containing hydrogen gas. When the temperature is below 80°C and there is no high tensile stress, the hydrogen absorption rate is very slow and no hydrogen embrittlement occurs. In pure hydrogen environment without water at high pressures and high temperatures, titanium has a serious tendency of hydrogen absorption and hydrogen embrittlement. The water content in hydrogen gas can significantly prevent the hydrogen absorption of titanium. Titanium reacts with pure nitrogen to form a golden yellow film above 540°C. At temperatures above 816°C, titanium becomes brittle due to the diffusion of nitrogen into it. More information about the reaction of titanium with elementary gas is described in Section 1.3.6.1 of Chapter 1.

Titanium can be corroded by hydrogen chloride gas. It reacts with dry hydrogen chloride and produces $TiCl_4$ at temperatures higher than 300°C (Huang et al 2009).

$$Ti + 4HCl \rightarrow TiCl_4 + 2H_2\uparrow \qquad \text{(reaction 2.9)}$$

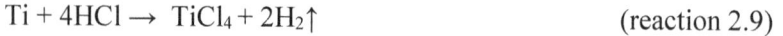

At room temperature, titanium has good corrosion resistance in dry sulfur dioxide, water saturated sulfur dioxide and water saturated hydrogen sulfide (Table 2-17). Titanium reacts with hydrogen sulfide and forms a protective film, but at elevated temperatures it reacts with hydrogen sulfide and produces hydrogen gas.

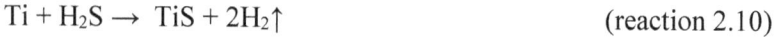

$$Ti + H_2S \rightarrow TiS + 2H_2\uparrow \qquad \text{(reaction 2.10)}$$

At 600°C powder titanium starts to react with hydrogen sulfide and produces sulfide, at 900°C and 1200°C the reaction product is mainly TiS and Ti_2S_3, respectively.

Titanium is resistant to corrosion of ammonia gas at room temperature. Above 150°C ammonia decomposes into nitrogen and hydrogen, titanium becomes brittle due to hydrogen absorption, the hydride on the surface spalls, so the corrosion rate of titanium in ammonia gas at 220°C can reach 11.2 mm/y (Zhu 1995).

Table 2-17 Corrosion rate of CP-Ti in sulfurous gases (After Huang and Zuo 2003)

Gas	Temperature (°C)	Corrosion rate (mm/y)
Dry sulfur dioxide	21	0
Water saturated sulfur dioxide	21	<0.003
Water saturated hydrogen sulfide	21	<0.127

2.2.3.7 Elevated Temperature Environments

Resistance of titanium to elevated temperature corrosion is determined by the oxide film on its surface and the aggressiveness of the medium. As a structural material in air or oxidizing atmosphere, titanium and its alloys can be used at temperatures up to 426°C. But hydrogen absorption becomes obvious at about 250°C. In hydrogen atmosphere above 316°C titanium becomes brittle due to hydrogen absorption. Hence, chemical equipment operating at temperatures above 330°C can not use titanium material if extensive tests have not been performed. As hydrogen absorption and mechanical property are a concern, the operating temperature of titanium pressure vessels should not exceed 250°C, and the upper limit of temperature of titanium tubes in heat exchangers is 316°C. Table 2-18 gives long term operating temperature for some titanium alloys (Huang et al 2009).

Table 2-18 Long term operating temperature for some titanium alloys

Category of titanium	Nominal composition	Operating temperature (°C)
CP-Ti	Ti	300
α-Ti	Ti-5Al-2.5Sn	500
	Ti-0.2Pd	350
near α-Ti	Ti-2Al-2.5Zr	350
	Ti-8Al-1Mo-1V	500
	Ti-3Al-2.5V	320
	Ti-6Al-2Sn-4Zr-2Mo-0.1Si	500
	Ti-1Al-1Mn	300
	Ti-2Al-1.5Mn	350
	Ti-4Al-1.5Mn	350
α-β-Ti	Ti-6Al-4V	400
	Ti-6Al-2.5Mo-1.5Cr-0.5Fe-0.3Si	450
	Ti-3Al-5Mo-4.5V	350
	Ti-5Al-2Sn-4Mo-4Cr	430
	Ti-5Al-4.75Mo-4.75V-1Cr-1Fe	400
	Ti-6Al-2Sn-4Zr-6Mo	400
β-Ti	Ti-15V-3Cr-3Sn-3Al	290
	Ti-10V-2Fe-3Al	320

2.2.3.8 Summary

Table 2-19 gives a summary of the anti-corriosn performances of titanium alloys in various service environments.

2.3 Localized Corrosion

One of the main failure forms of titanium equipment is localized corrosion.

2.3.1 Pitting Corrosion

Pitting corrosion is a kind of localized corrosion, which results in pits in the metal surface. Titanium alloys have better pitting corrosion resistance than stainless steel, aluminum alloys and nickel-base alloys. At elevated temperatures if the water content is below a critical value, pitting corrosion occurs on titanium in high concentration chloride solutions and some organic media, such as glacial acetic acid, formic acid, methyl alcohol and ethyl alcohol containing inorganic salt (Huang et al 2009).

Table 2-19 Anti-corriosn performances of titanium alloys in various service environments

Medium		Good anticorrosion	Bad anticorrosion
Atmosphere		All atmospheric environments	—
Marine		Neutral seawater below 130°C	Seawater with high temperatures or low pH
Inorganic acids	HNO$_3$	Temperature (T)<70°C	T>70°C
	HCl	Concentration (C)<7%, 25°C C<0.5%, 100°C	C >7%, 25°C C>0.5%, 100°C
	H$_2$SO$_4$	C <20%, 0°C C <5%, 25°C C<0.2%, 100°C	C >20%, 0°C C >5%, 25°C C>0.2%, 100°C
	H$_3$PO$_4$	C<30%, 25°C C <10%, 60°C C <2%, 100°C	C>30%, 25°C C >10%, 60°C C >2%, 100°C
	HF	—	All concentrations
Organic acids		Most organic acids, such as acetic acid, citric acid, tartaric acid, lactic acid and tannic acid	Hot concentrated oxalic acid, formic acid, trichloroacetic acid and sulfamic acid (in deaerated condition
Organic compounds		Alcohols, aldehydes and hydrocarbons, with air or moisture	No air and water
Alkalis		All concentrations in all alkaline solutions (pH<12, below boiling point)	pH>12, boiling (hydrogen absorption)
Salts		Most of salt solutions (chloride, oxidizing salts)	AlF$_3$, concentrated AlCl$_3$, ZnCl$_3$, CaCl$_2$ (pitting corrosion, crevice corrosion)
Gases	Cl$_2$	Wet	Dry
	O$_2$	<500°C	>500°C
	N$_2$	<700°C	>700°C
	H$_2$	<500°C	>500°C
	Halogen gas	—	All halogen gas
	SO$_2$	Dry, water saturated, 25°C	Elevated temperature
	NH$_3$	<150°C	>150°C
	HCl	<300°C	>300°C
	H$_2$S	<600°C	>600°C

Pitting corrosion readily takes place where there are crevices. For example, when titanium exchanger is used in zinc chloride solution, pitting corrosion easily develops where titanium is coupled with iron. In addition, pitting corrosion is often observed in color-changing areas caused by welding or inappropriate heat treatment, as well as iron-polluted areas.

Pitting corrosion resistance of titanium alloys is influenced by not only material properties but also environmental factors, consisting of the medium, pH value, temperature. Some effects considering metallurgy and environment are described below:

1) CP-Ti contains TiFe, which often becomes the nucleation site of pitting corrosion. Thus, the pitting corrosion resistance decreases with an increase in iron content, but increases with an increase in oxygen content.

2) Temperature rise may decrease pitting corrosion resistance in chloride and bromide solutions. The pitting sensitivity of titanium increases with an increase in halide concentration, and decreases with an increase in flow rate.

3) Surface treatment plays an important role in pitting initiation of titanium. Table 2-20 shows the effect of four surface treatment methods on pitting potential of titanium in boiling 1% NaBr solution. Titanium polished with wet sandpaper has the largest pitting sensitivity. In this table, there are three grades of CP-Ti. The contents of iron and oxygen for grade 1, 2 and 3 are 0.0253% and 0.0655%, 0.0440% and 0.0955%, 0.1440% and 0.2145%, respectively.

Table 2-20 Effect of surface treatment methods on pitting potential of CP-Ti (boiling 1% NaBr, pH=6) (After Wang and Tian 2007)

Surface treatment method	Pitting potential vs. SCE (V)		
	Grade 1	Grade 2	Grade 3
Polishing using 400 wet sandpaper	1.05	1.16	1.23
Vacuum annealing, 700°C/10 min	1.21	1.37	1.43
Anodizing, 2%H$_2$SO$_4$, 2 V/10 min	1.24	1.39	1.42
Thermal oxidation in air, 600°C/10 min	1.16	1.30	1.35

4) Pitting corrosion is easily initiated on rough surface, or surfaces rubbed with zinc, iron, aluminum, manganese or copper. Some anions, such as SO_4^{2-}, NO_3^-, CrO_4^{2-}, PO_4^{3-} and CO_3^{2-}, help to improve the pitting corrosion resistance.

2.3.2 Crevice Corrosion

This is a localized form of corrosion caused by the existence of crevices between adjoining surfaces. Due to its excellent corrosion resistance in chloride solutions, titanium has been widely used in oceanic parts and chemical equipment. But when there are crevices existing between adjoining surfaces, titanium suffers from serious crevice corrosion, causing great damage.

Factors affecting crevice corrosion of titanium include temperature, chloride concentration, pH, geometry of crevice, etc. The crevice corrosion sensitivity increases with an increase in solution concentration and temperature, as well as a decrease in pH. The test results in Table 2-21 indicate that in supersaturated sodium chloride solution, the incubation period of crevice corrosion of titanium is prolonged by increasing pH value and lowering temperature. The geometry of the crevice and the area ratio of exterior to interior crevice have an impact on crevice corrosion. The corrosion rate in wide crevices is much slower than in narrow crevices. In addition, crevice corrosion has a greater tendency to occur in crevices between titanium and a non-metal, such as Ti-Teflon, Ti-carbon fluoride, Ti-silicone rubber and Ti-asbestos, than in crevices between Ti and Ti (Zhang et al 2005).

Table 2-21 Effect of pH and temperature on crevice corrosion incubation period of titanium ("+"----presence of crevice corrosion, "–"----no crevice corrosion) (After Huang et al 2009)

Medium	Test time (h)	Temperature (°C)							
		110	120	130	140	150	160	180	200
270g/L NaCl + 1g/L HCl	48	–	–	–	+	+	+	+	+
	96	–		+	+	+	+	+	+
	200			+	+	+	+	+	+
	800				+	+	+	+	+
270g/L NaCl + 0.05g/LNaOH	48	–	–	–	–	–	+	+	+
	96	–	–	+	+	+	+	+	+
	200			+	+	+	+	+	+
	800				+	+	+	+	+

2.3.3 Galvanic Corrosion

Galvanic corrosion occurs when two dissimilar metals with different electrochemical potentials are coupled in a corrosive electrolyte. When titanium is coupled with other metals, such as zinc, aluminum, copper, it has a very positive potential, hence, it acts as a cathode and absorbs hydrogen. On the other hand, the coupling metal corrodes fast as an anode. Therefore, this kind of galvanic corrosion should be prevented. Table 2-22 gives a galvanic series of metals in flowing seawater (3.96 m/s, 23.9°C). According to the galvanic series, a metal with more negative potential in contrast with titanium tends to corrode when connected to titanium.

Under certain circumstances hydrogen is released and absorbed by titanium. This may cause hydrogen embrittlement at elevated temperatures. Therefore, hydrogen absorption should be paid attention and prevented on titanium in the presence of galvanic effect.

Table 2-22 Galvanic series of metals in flowing seawater (3.96 m/s, 23.9°C) (Sources: Zhu 1995, Ahmad 2006)

Material	E vs. SCE (V)
Platinum	+0.15
Zirconium	-0.04
18-10 stainless steel (passive)	-0.05
18-8 stainless steel (passive)	-0.08
Monel alloy 400	-0.08
Hastelloy alloy C	-0.08
Titanium	-0.10
Silver	-0.13
18-10 stain steel (active)	-0.18
Nickel	-0.20
18-0 stainless steel (passive)	-0.22
70-30 Copper nickel	-0.25
80-20 Copper nickel	-0.27
90-10 Copper nickel	-0.28
Naval brass	-0.29
Aluminum brass	-0.32
Copper	-0.36
18-8 SS steel (active)	-0.53
18-0 stainless steel (active))	-0.56
Carbon steel	-0.61
Cast iron	-0.61
Aluminum	-0.79
Zinc	-1.03

In a coastal power plant (Huang et al 2009), titanium equipment was coupled with 8 mm thick carbon steel pipes. The pipes experienced corrosion failure after 1.5 months, which severely influenced safety in production. Galvanic corrosion rate depends not only on their electrochemical potentials, but also on their surface areas and polarization characteristics. The environment for galvanic corrosion of titanium can fall into the following two types according to different media:

Type I, titanium has good corrosion resistance in the media, such as saline solution, nitric acid, acetic acid, seawater, tap water and atmosphere, etc. In this scenario, titanium is not affected while the coupled metal suffers more serious corrosion.

Type II, in some media, such as hydrochloric acid, sulfuric acid, oxalic acid, etc, titanium is either activated or passivated. In both cases, the corrosion rate of titanium and the coupling metal may be affected.

In practical marine engineering, normally the composite structure consists of titanium alloy and other different materials. Test results on galvanic corrosion between Ti-6.5Al-2Zr-1Mo-1V and other metals (cast steel, alloy steel aluminum alloy and copper alloy) in static seawater are as follows (Huang et al 2009):

1) When cast steel or nickel copper alloy was coupled with Ti-6.5Al-2Zr-1Mo-1V, the galvanic corrosion rate increased with an increase in area ratio. Galvanic corrosion rate of cast steel was accelerated because of general corrosion, and that of nickel copper alloy, which was 1-2 order of magnitudes lower than cast steel, was increased due to pitting corrosion.

2) Galvanic corrosion rate of red copper and brass was small. Increasing the surface area of Ti-6.5Al-2Zr-1Mo-1V may promote passivation of red copper, which accordingly impeded the galvanic corrosion.

3) When the area ratio was 1:1, corrosion of cast steel was obviously accelerated, but that of red copper, silicon brass, high manganese aluminium bronze and nickel copper alloy remained almost unchanged.

4) Galvanic corrosion between Ti-6.5Al-2Zr-1Mo-1V and high chromium steel (05Cr25Ni17MoCuN) was small and negligible (10^{-4} mm/y).

5) Galvanic corrosion between Ti-6.5Al-2Zr-1Mo-1V and aluminum alloy (Al-4.5Cu-1.1Li-0.5Mn-0.2Cd, Al-5.6Zn-2.5Mg-1.6Cu-0.26Cr), or steel (30CrMnSiA, 30CrMnSiNi2A, 16CrNiSi) was obvious when the temperature was above 300°C.

2.4 Hydrogen Embrittlement

A limitation in the application of titanium alloys is their sensitivity to hydrogen embrittlement (Madina and Azkarate 2009, Lin and Du 2014), which is an important factor for the failure of titanium facilities.

2.4.1 Hydride Formation

Titanium is a very reactive metal, and it is very easy to absorb hydrogen. When the hydrogen content is (80-150) ppm, a needle-like hydride phase can be observed using optical microscopy. With the further increase of the hydrogen content, the number of hydride increases. This can lead to a large volume expansion. There are several situations for the corrosion damage of titanium caused by hydride formation.

1) If the hydrogen diffusion rate is slow, the hydride is mainly concentrated on the titanium surface, then the surface hydride becomes brittle and peels off, resulting in the acceleration of corrosion;

2) Hydrogen diffuses to the stress-concentrated position in the stress field and produces hydride, causing the formation of internal cracks. This is hydrogen-induced cracking;

3) Significant amounts of hydrogen are absorbed in the titanium substrate, leading to hydrogen embrittlement.

2.4.2 Influencing Factors

1) Hydrogen and water content in hydrogen atmosphere

Hydrogen is a direct cause for hydrogen embrittlement. At room temperature, its solid solubility is 20-200 mg/kg in titanium. When its content is in excess of the solid solubility limit, TiHx begins to appear. Therefore, susceptibility to hydrogen embrittlement is decreased by increasing the solid solubility. Ductility of high purity titanium is reduced by 9% if the content of hydrogen is 500 mg/kg, and ductility loss is 23% for CP-Ti. If the hydrogen content keeps rising, the ductility drops sharply. In conclusion, susceptibility to hydrogen embrittlement of titanium increases with an increase in hydrogen content in a certain range.

Generally, hydrogen embrittlement develops when the hydrogen content in titanium part is above the solid solubility limit. Allowable hydrogen content can be expressed with the following equation (Zhang et al 2005):

$$C = C_{\text{limit}} / n \qquad\qquad (2.1)$$

where C is allowable hydrogen content; C_{limit} is solid solution limit; n is determined by the importance of structural material, service conditions and manufacturing process. For titanium parts without welding and high stress concentration, $n = 1.5$-2; for un-annealed welding structure, n = 2.5-3.5; for annealed welding structure, $n = 2$. Table 2-23 shows the minimum hydrogen content causing hydrogen embrittlement for some titanium alloys.

Table 2-23 Minimum hydrogen content causing hydrogen embrittlement for some titanium alloys (After Zhang et al 2005)

Type of titanium	Hydrogen content (ppm)
Ti-6Al-4V	8
Ti-8Al-1Mo-1V	5
Ti-2Fe-2Cr-2Mo	10
Ti-4Mo	20
Ti-4Al -3Mo-1V	26
T i-6Al-6Mo-2Sn	38
T i-4Al-1.5Mo-5V	30

Water content in hydrogen atmosphere plays an important role in hydrogen absorption of titanium.

The influence of the water content on hydrogen absorption of titanium in hydrogen atmosphere with a pressure of 5.5 MPa at 316°C is given in Table 2-24.

Table 2-24 Influence of water content on hydrogen absorption of titanium in hydrogen atmosphere with a pressure of 5.5MPa at 316°C (After Wang and Tian 2007)

Water content (mass fraction, %)	Hydrogen absorption (ppm)
0	4486
0.5	5100
1.0	698
2.0	7
3.3	10
5.3	17
10.2	11
22.5	0
37.5	0
56.2	0

The maximum hydrogen absorption occurs in dry hydrogen atmosphere, and adding a certain amount of water (2%) obviously lowers the hydrogen absorption. Water contains oxygen, which keeps the oxide film on titanium surface in good condition, even repairs the damaged film and then prevents hydrogen penetration. The reason why titanium has been widely and safely used in oil refining processes containing a large amount of hydrogen is partly because there is water. It can be inferred that oxygen in a hydrogen atmosphere can prevent hydrogen absorption. Therefore hydrogen absorption of titanium in air is much slower than in pure hydrogen.

2) Environmental factors

Temperature

Temperature rise not only increases the reaction rate between titanium and hydrogen, but also accelerates hydrogen diffusion in α-Ti. When temperature is higher than 80°C, obvious hydrogen absorption can be detected; when temperature exceeds 300°C, the reaction rate is increased and lots of hydride is produced in titanium. Hydrogen diffusion has a decisive role in hydrogen embrittlement of titanium, which rarely occurs under 80°C because hydrogen diffusion is slow at low temperatures. Generally, the relationship between the hydrogen diffusion coefficient in metal and temperature can be described by Eq.(2.2) (Zhang et al 2005):

$$D = D_0 \exp(\frac{Q_d}{RT}) \tag{2.2}$$

where D is diffusion coefficient (m^2/s), Q_d is the diffusion activation energy (J/mol), T is temperature, D_o is diffusion constant (m^2/s), R is gas constant (8.314 J·mol^{-1}·K^{-1}).

Interstitial atoms in body-centered cubic (bcc) structure have the least diffusion activation energy, and the energy of atoms is much higher in a fcc and hcp structure. During transformation from bcc β-Ti to hcp α-Ti, Q_d is increased from 2.8×10^7 kJ/gram molecule to 5.3×10^7 kJ/gram molecule. Apparently, hydrogen diffusion rate in α-Ti is much slower than in β-Ti. Temperature also has a distinct influence on hydrogen absorption and diffusion in titanium. The hydrogen absorption increases with a rise in temperature, and the kinetic curves of hydrogen absorption obey parabolic law at low temperatures and linear regularity at high temperatures.

Electrochemical potential

There is a critical potential for hydrogen absorption of titanium in electrolyte solution. At room temperature, if corrosion potential is more negative than the critical potential, hydrogen depolarization reaction begins, and absorbed hydrogen accumulates until hydrogen embrittlement takes place. An attempt has been made to calculate the potential at which titanium hydride was produced according to its free energy of formation. Suppose there are two possible reactions as follows:

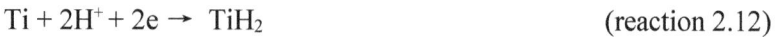

$$Ti + 1.75H^+ + 1.75e \rightarrow TiH_{1.75} \qquad \text{(reaction 2.11)}$$

$$Ti + 2H^+ + 2e \rightarrow TiH_2 \qquad \text{(reaction 2.12)}$$

The potential at which titanium hydride is produced is

$$E1 = 0.008 - 0.059pH \ (V, \ vs. \ SCE) \qquad (2.3)$$

$$E2 = -0.226 - 0.059pH \ (V, \ vs. \ SCE) \qquad (2.4)$$

pH in seawater is 8.2, but reaches 11 in the electrolyte near the cathode at low current density and 12 at high current density because OH^- is produced. Then, the potential is:

$$E1 = -0.642 \ (V, \ SCE) \ (pH=11)$$

$$E2 = -0.875 \ (V, \ SCE) \ (pH=11)$$

$E1$ is closer to the critical potential, and therefore the first reaction is more likely to occur. If the second reaction also occurs, then more negative potential is needed.

Long term test results showed that in neutral salt solutions the cathodic potential which was lower than -0.7 V induced hydrogen absorption. Very high cathodic current density (with more negative potential than -1.0 V vs. SCE) accelerated hydrogen absorption and eventually caused hydrogen embrittlement.

Medium

Titanium shows different hydrogen absorption behavior in different medium. Table 2-25 lists the rate of hydrogen absorption of titanium in acids. Hydrogen absorption takes place in several non-oxidizing acids, because titanium is in the active corrosion state, and the hydrogen absorption is caused by reduction of the main oxidant H^+. However, in the oxidizing nitric acid, NO_3^- has a stronger reduction capability than H^+ and thus hydrogen absorption does not occur.

Table 2-25 Hydrogen absorption of titanium in different acids (After Zhang et al 2005)

Test condition	Acid	Corrosion rate (mg/cm^2)	Rate of hydrogen absorption (mL/cm^2)	Hydrogen absorption (%)
35°C, 240h	10% H_2SO_4	12.0	1.3	16
		11.0	1.3	17
	10% HCl	9.0	1.6	25
		8.8	1.6	27
	10% $H_2C_2O_4$	20.0	0.41	2.9
		23.0	0.33	2.1
	1% H_2SO_4	3.9	1.3	49
		3.4	1.2	49
	3% HCl	3.2	0.64	29
		3.3	0.52	23
Boiling, 6h	0.2% $H_2C_2O_4$	6.4	0.72	16
		6.4	0.69	15
	65% HNO_3	0.15	0.00	—
		0.16	0.00	—

pH

Oxide film on the surface of titanium is stable in a medium with a pH range of 3 to 12, and it is a significant barrier to prevent hydrogen permeation. In a medium within the above pH range, no hydrogen absorption was found in short term cathodic hydrogen charging tests. If pH is greater than 12 or lower than 3, the oxide film is not stable, the substrate is not effectively protected. This facilitates hydrogen permeation.

Oxidizing agent and inhibitor

Hydrogen absorption of titanium is the result of hydrogen evolution in the cathodic reaction. If an oxidizing agent exists, which captures electrons and takes the place of hydrogen evolution on the cathode, no hydrogen absorption occurs. For example, nitric acid is added in hydrofluoric acid to reduce hydrogen absorption in acid pickling of titanium. Besides nitric acid, heavy metal ions, oxygen, some oxidizing anions, organic inhibitors have an inhibition effect for titanium corrosion, which simultaneously reduces hydrogen absorption.

3) Stress

Stress is also an important factor to promote the enrichment of titanium hydride and increase hydrogen diffusion, both of which increase susceptibility to hydrogen embrittlement. Generally, when there is not much hydride in titanium, external stress or internal residual stress cause their enrichment in the stress area, and then reduce the local plasticity, which finally result in hydrogen embrittlement. Meanwhile, stress accelerates hydrogen diffusion in titanium and also increases the risk of hydrogen embrittlement. The hydrogen diffusion rate increases with a stress increase in the range of 0 to 5.7 MPa.

4) Material factors

Alloying element and impurity

It has been found that some alloying elements and impurities have an influence on hydrogen absorption in titanium. Ni and Fe promote hydrogen absorption, Mo, Al, O tend to reduce it. But increasing Al or O content promotes stress corrosion of titanium. The effect of Pd is complicated, as it mainly depends on the corrosion medium.

Surface condition

Oxide film acts as a barrier on the titanium surface. The importance is the thickness and performance of the oxide film. The film composition varies with the environment, generally composed of TiO_2, Ti_2O_3 and TiO. It has semiconductor properties, and there are O^- vacancies. Increasing oxygen absorption reduces the defects on the film, and then prevents the permeation of harmful ions and hydrogen. Therefore, the measures in which titanium corrosion potential is moved toward positive direction and oxidizing film is stabilized are good for preventing hydrogen absorption.

Titanium with anodic oxidation film or thermal oxidation film on the surface has the best resistance to hydrogen absorption and embrittlement; then the acid pickled (nitric and hydrofluoric acid) or annealed surface; mechanical ground or polished surface comes last. It indicates that active titanium surfaces easily absorb hydrogen, whereas titanium with intact oxidation film has good resistance to hydrogen absorption and embrittlement (Zhu 1995).

If titanium is abraded by iron, some iron particles are left on its surface and become corrosion initiation points. In electrolyte solution the iron contaminants promote hydrogen absorption, especially in welding areas. Normally, no hydrogen evolution occurs in oxidizing solutions, even when oxidability is very weak. But it occurs in the moderate oxidizing solution when iron pollution is present because the potential in the local area on titanium surface drops below the hydrogen evolution potential. In the production process of urea, ethylene, acetate or terephthalic acid, as well as sulfuric acid, hydrochloric acid, nitric acid, phosphoric acid, formic acid, acetic acid, pypocholoride, chlorine dioxide, iron pollution must be paid attention where titanium is used (Zhu 1995).

The effect of iron contaminants and anodic oxidation on titanium depends on operating conditions. In some media iron pollution cannot always cause hydrogen absorption of titanium which is also related to oxidation properties of the medium, as well as contamination level, contamination depth, solution flowing rate, temperature, etc. In strong oxidizing medium, the corrosion potential of titanium in local areas contaminated by iron is still higher than the hydrogen evolution potential, and there is no hydrogen evolution. Also, hydrogen evolution does not lead to hydrogen embrittlement of titanium due to low hydrogen diffusion when the temperature is lower than 80°C. In addition, a solution with high flow rate may wash out iron contaminants and then reduce hydrogen evolution. It should be noticed that anodic oxidation film can be destroyed by strong reducing agents. Under this circumstance hydrogen absorption is inevitable, so some measures are needed to prevent it.

Material microstructure

When titanium absorbs hydrogen and its content reaches the limit of solid solubility, TiHx forms. α-Ti and β-Ti have different solid solubilities of hydrogen, as shown in Table 2-26. α phase has a hcp structure, while β phase has a bcc structure which easily absorbs hydrogen. At 300°C, the solid solubility limit of β-Ti is equal to 5.5 times that of α-Ti, therefore hydrogen absorption in α-β-Ti or β-Ti is much more than that in α-Ti.

Table 2-26 Solid solubility of hydrogen in different microstructure of titanium (After Wang and Tian 2007)

Microstructure	Temperature (°C)	Atom fraction (%)	Mass fraction (%)
α	room temperature	0.1	0.002
α	300	8	0.18
β	300	44	0.99

2.5 Stress-corrosion Cracking

Stress-corrosion Cracking (SCC) is the cracking induced from the combined influence of tensile stress and a corrosive environment. It can lead to the unexpected sudden failure of metals subjected to a tensile stress.

Titanium is susceptible to SCC in such media as fuming nitric acid, methyl alcohol solution, hydrochloric acid, pypocholoride at high temperatures, molten salt at 300-450°C, atmosphere containing NaCl, carbon disulfide, normal hexane and dry chlorine. SCC susceptibility of titanium in nitric acid increases with increasing NO_2 content and decreasing water, and reaches a maximum in pure nitric acid containing 20% NO_2. In concentrated nitric acid titanium experiences SCC at room temperature when the NO_2 content is higher than 6.0% and H_2O is lower than 0.7%. CP-Ti is susceptible to SCC in 10% hydrochloric acid and 0.4% hydrochloric acid-methanol solution. Although titanium suffers from SCC in some specific media, it has better SCC resistance than most other metals.

2.5.1 Mechanisms

Generally, a consistent view is that mechanisms of SCC are classified as anodic dissolution and hydrogen induced cracking. Recent research on anodic dissolution SCC shows that the formation and maintenance of the surface passive layer or loose layer induces an additional high tensile stress, so dislocation appears and moves under a very low applied stress. When local plastic deformation promoted by corrosion reaches the critical state, stress concentration at certain points (dislocation-free zone or the front of dislocation pile-up group) is equal to the atomic binding force, and then SCC microcrack nucleation is induced. The corrosion of medium prompts the microcrack to propagate into a cleavage or along the grain boundaries rather than blunting into a void, and finally leads to brittle fracture under a low stress. Hydrogen induced SCC involves the following processes:

- H^+ migrates and discharges;
- Some of the hydrogen atoms that absorb on the titanium surface stick together to form hydrogen molecules and then escape as bubbles; and
- Other hydrogen atoms dissolve into the metal, diffuse and then aggregate in the stress concentration zones, finally resulting in brittle fracture under a low stress.

Mechanisms of SCC of titanium vary with alloy type and the medium. In summary, there are three kinds of SCC mechanisms for titanium: active channel cracking, hydrogen embrittlement and chlorine embrittlement. It should be pointed out that the these three mechanisms are not independent, but are inseparably interconnected (Zhang et al 2005).

2.5.2 Influencing Factors

SCC is the conjoint action of the material, corrosive environment and stress. If any one of these factors is eliminated, SCC initiation becomes impossible (Zhang et al 2005).

1) Environment

Medium

SCC of titanium may occur in aqueous solutions, organic solutions and hot salts. For example, methyl alcohol-chloride solution, fuming nitric acid, nitrogen tetroxide, NaCl solution, as well as hot chloride (>290°C), methyl alcohol, hydrochloric acid, seawater, nitric acid, organic acid, molten salt, liquid state N_2O_4, carbon tetrachloride, hydrogen, bromine vapour, etc. Contacting with cadmium may also cause SCC of titanium.

Sometimes minor impurities are also a major factor in SCC, for example, trace amounts of water are very influential in SCC of titanium in methyl alcohol. If the water content is above 0.2%, no SCC of Ti-6Al-4V occurs in methyl alcohol+$0.1MH_2SO_4$ solution. Conversely it happens when the water content is under 0.2%.

pH

There are different opinions about the effect of pH on SCC of titanium. Generally, SCC susceptibility decreases with increasing pH, and SCC is inhibited when the pH is between 13 and 14.

Electrochemical potential

As an electrochemistry parameter, electric potential (E) determines the possibility of an electrochemical reaction from thermodynamics, and may play a crucial role in SCC. For material that is susceptible to SCC, the potential range of susceptibility increase corresponds to thermodynamic unstable regions in a E-pH diagram. Besides, when SCC of titanium occurs, the pH value of medium in the crack tip is decreased to 1.7-1.8. Therefore, corrosion behavior of the crack tip, that is, SCC susceptibility is determined by potential and pH. Local environment with specific electrochemical characteristics is a prerequisite for SCC to occur. Analysis of potential-time and polarization curves can help understand crack initiation, growth and growth rate.

Different corrosive systems have different critical potential and susceptible zones of SCC. For example, SCC of β-Ti alloy in halide solution aggravates when the potential is about -600 mV; cracks occur at transpassive potential, but no cracks occur when potential is under -1000 mV. Susceptible potential region of SCC for Ti-8Al-Mo-V in solutions containing Cl⁻ and Br⁻ is -500~-600 mV, but in solutions containing I⁻, this region is above 0 mV. Table 2-27 shows the breakdown potential of SCC for some titanium alloys.

Table 2-27 Breakdown potential of SCC for some titanium alloys (After Zhang et al 2005)

Type of titanium	Medium	Breakdown potential vs. SCE (V)
Ti-8Al-Mo-V	0.6M KCl	-0.8~0
Ti-8Al-Mo-V	0.6 M NaCl, room temperature	-0.76
Ti-7Al-2Nb-1Ta	3% NaCl, room temperature	-1.1

Temperature

Temperature is one of the key factors for SCC. Generally, SCC susceptibility increases with the temperature. Hot salt stress corrosion (HSSC) of Ti-6Al-13Mo-2Zr-0.5Sn occurs when temperature is above 450°C.

2) Stress

One of the essential conditions for SCC is the presence of tensile stress. 40% of SCC failures were caused by residual stress induced from cold machining, forging, welding, heat treatment or assembling. Tensile stress has many sources, which are mainly divided into two types:

a)   processing and manufacturing, for example, welding, mechanical machining, heat treatment, etc.

b)   assembly and operation, for example, tightening screws, periodic thermal cycling, load, pressure, etc.

Of all the residual stress sources, welding may be the most common cause of the SCC failures. Most SCC cracks of equipment appear in the area where welding stress concentrates and reaches the yield strength of the material. Generally, local stress that causes cracks must exceed the yield strength. In addition, inhomogeneous stress caused by volume effects of corrosion products is also the stress source of SCC. The incubation period of SCC decreases with increasing stress. Residual stresses can be relieved by stress-relief annealing. However, this must be done in a controlled way to avoid creating new regions with high residual stress, and its effect on material strength, plasticity and toughness must be considered.

3) Material

In the same medium, SCC behavior of titanium alloys varies with different chemical components, segregation, grain size, crystal defects, heat treatment, surface conditions, etc.

α-Ti alloy is more susceptible to SCC than other titanium alloys. Al content in α-Ti alloy has a significant impact on K_{ISCC} (SCC fracture toughness), whereas the impact on K_{Ic} (fracture toughness) is not obvious. Titanium alloy containing a 6.5-8% Al has the lowest $K_{ISCC.}$ There is little difference between K_I and K_{ISCC} when the Al content is lower than 5.5%. It is generally accepted that a 6%Al in titanium alloy is the threshold for SCC to occur in salt solutions. The reason is that an ordered α2 phase is formed when the Al content is above 5-6%, its existence changes the dislocation substructures caused by deformation, consequently the local plane slips and partial oxide film ruptures on the surface, and then the titanium substrate is exposed to the corrosive medium. It was found that low temperature aging treatment did not affect the fracture toughness of Ti-4Al, but had an obvious impact on the fracture toughness of Ti-8Al (Zhang et al 2005). This is because lots of ordered α2 phase was produced in Ti-8Al after low temperature aging, which greatly lowered K_I and K_{ISCC}, and increased the SCC rate. Sn additions in α-Ti alloy lower its SCC resistance. It promotes the alloy to produce solid solution containing Sn after quenching and aging, increasing phase boundaries. This explains why the ternary alloys suffers from SCC after low temperature aging.

Adding Mo and V, which are conducive to the formation of β phase, minimizes SCC susceptibility, because β phase can prevent crack growth in α phase for α+β alloy. The effect of microstructure after different heat treatment on fracture toughness of Ti-8Al-Mo-V is given in Table 2-28. It shows clearly that titanium alloy with acicular martensite has higher K_I and K_{ISCC} values. In addition, adding a small number of Pb and Ru also lowers the SCC susceptibility.

Titanium alloys which are susceptible to hydrogen embrittlement have higher SCC susceptibility, because hydrogen diffuses to the crack tip and forms titanium hydride with the lamellar structure, favoring SCC growth.

2.6 Material Selection and Use Design

2.6.1 Material Selection

Corrosion resistance and service reliability of titanium alloys must be considered in material selection (Lin and Du 2014). Generally, CP-Ti is used in oxidizing media, and corrosion resistant titanium alloys are used

Table 2-28 Effect of microstructure after different heat treatment on fracture toughness of Ti-8Al-Mo-V (After Zhang et al 2005)

Heat treatment	Microstructure	Medium	K_I (N/mm$^{3/2}$)	K_{SCC} (N/mm$^{3/2}$)
Double annealing	Equiaxial α+β	Argon gas protection 3%NaCl	1510	618-686
870°C 2h	Equiaxial α+β	Argon gas protection 3%NaCl	2677	686-735
927°C 2h	Equiaxial α+β	Argon gas protection 3%NaCl	2814	1030-1108
982°C 2h	Acicular martensite	Argon gas protection 3%NaCl	2952	2883
109°C 2h	Acicular martensite	Argon gas protection 3%NaCl	2991	2952

in non-oxidizing media (neutral or reducing media). For example, in medium where crevice corrosion is most likely to occur, using Ti-0.2Pd or Ti-0.3Mo-0.8Ni is recommended. Ti-5Ta is recommended in strongly reducing media (such as high concentration nitric acid at elevated temperature). Ti-32Mo, as a high alloying β-type titanium alloy, is used in high concentration sulfuric acid due to its excellent corrosion resistance. But it has a poor processing performance, and segregation can easily occur during smelting.

CP-Ti, Ti-2.5Cu, Ti-2Al-1.5Mn and Ti-4Al-1.5Mn are commonly selected to make complicated sheet parts operating under small stress and a temperature of 300°C; Ti-6Al-4V is widely used to produce force-bearing parts operating under 400°C; Ti-6.5Al-1.5Zr-3.5Mo-0.3Si is used for parts operating under 500°C; Ti-5Al-2.5Sn and Ti-6.5Al-2Zr-1Mo-1V are used for welding structures under a large stress. Titanium alloys operating at 600°C is now under investigation. If operating temperature reaches 700°C, Ti$_3$Al intermetallic compound is used and at 900°C TiAl intermetallic compound is considered.

2.6.2 Practical Use Design

Some important points to prevent corrosion in practical use of titanium alloys need to be considered (Huang et al 2009).

(1) Avoid using in these four inorganic acids (hydrofluoric acid, hydrochloric acid, sulfuric acid, and orthophosphoric acid), these four hot high concentration organic acids (oxalic acid, formic acid, trichloroacetic acid and sulfamic acid), as well as highly corrosive aluminium chloride solution. These media may cause serious corrosion on titanium alloys. However, some oxidizing agents, such as heavy metal ions or nitric acid, can impede the corrosion.

(2) Avoid using in fuming nitric acid.

(3) Avoid using in dry chlorine. Titanium reacts with dry chlorine and produces $TiCl_4$, releasing a lot of heat, and hence cannot be used in dry chlorine containing 0.1-0.3% water.

(4) Avoid using in liquid oxygen or water solution with a high oxygen partial pressure. Titanium is sensitive to impact or friction in liquid oxygen. Spontaneous combustion occurs on a fresh titanium surface (without oxidation film) at room temperature and a pressure of 0.35 MPa.

(5) Avoid iron pollution on the titanium surface. The presence of iron promotes hydrogen absorption and increases the risk of hydrogen embrittlement.

(6) Avoid using in a high-temperature condition. For example, the operating temperature of CP-Ti should be lower than 250°C; in an oxidizing atmosphere the operating temperature of this material should be lower than 450°C.

(7) Avoid crevice and recess to the largest extent in structural design. For example, welding is used instead of a bolted joint; butt welding replaces spot welded lap joints.

(8) Pay attention to thermal stress caused by different coefficients of thermal expansion when titanium is coupled with steel, which may cause structure fracture.

(9) Titanium alloys are sensitive to hot-NaCl stress corrosion, and hence non-chlorine solvent cleaner should be used to deal with titanium parts at a temperature above 230°C.

2.7 Prevention Measures for Corrosion

Titanium alloys may be subject to general corrosion, pitting corrosion, hydrogen embrittlement, crevice corrosion and stress corrosion in some cases. This section gives a summary of prevention measures corresponding to different forms of corrosion for titanium alloys (Lin and Du 2014, Bardal 2004)

2.7.1 Pitting Corrosion

(1) Avoid using titanium as structural material in a strong oxidizing acidic medium at elevated temperatures, and critical conditions for the pitting corrosion should be determined through laboratory experiments.

(2) Apply galvanic anodic protection on titanium alloys in solutions containing halide ions. Titanium cannot be used as an insoluble anode in electrochemical protection for oceanic parts.

(3) Avoid pollution of zinc, iron, aluminum, manganese and steel in acid pickling, chemical polishing, elevated temperature vacuum annealing or anodic oxidation.

(4) Increase flow rate of medium containing chloride.

(5) Add an inhibitor, such as sulfate.

2.7.2 Crevice Corrosion

To prevent titanium parts from crevice corrosion, choose a proper titanium alloy, optimize the structural design, use a special gasket, or improve the corrosion resistance by surface treatments.

(1) Eliminate or minimize crevices and stagnant areas at the design stage. For example, use welded joints instead of bolted or riveted joints; use butt joints instead of spot welded joints; use arc components instead of flat components.

(2) Use surface treatment, including palladium coating, thermal oxidation, chemical oxidation, anodizing in special medium, etc.

(3) Choose a proper titanium alloy and improve the material compatibility. Ti-0.2Pd, Ti-0.8Ni-0.3Mo and Ti-32Mo have been developed to improve the crevice corrosion resistance of titanium alloys, and Ti-0.2Pd has the best performance. No crevice corrosion was observed on Ti-0.15Pd in 6%NaCl solution at 125°C.

(4) Use a special gasket. To seal titanium surface, some special gaskets have been developed to eliminate the crevice corrosion.

2.7.3 Galvanic Corrosion

Under ordinary atmosphere titanium alloy is allowed to contact stainless steel or nickel base alloy without surface protection, but in oceanic climate direct contact with electronegative material should be forbidden when the temperature is above 300°C. Measures such as protective coating should be taken to protect the titanium parts. Electronegative material (such as aluminum alloy or constructional steel) that contacts titanium act as an anode and suffer serious corrosion. Insulation connection or reducing the potential difference is often used to prevent the galvanic corrosion.

(1) Anodic oxidation of an aluminum alloy can reduce galvanic corrosion between Ti-6.5Al-2Zr-1Mo-1V and the aluminum alloy, but the corrosion cannot be totally prevented unless coating is further applied on the oxidized aluminum alloy;

(2) Surface treatment and protective primer on steel are effective methods to mitigate galvanic corrosion between titanium and steel;

(3) Coating or anodizing on titanium parts is further employed to inhibit galvanic corrosion between titanium and the coupled metal;

(4) Anti-corrosion adhesive tape should be used between titanium and electronegative materials (such as aluminum alloy or constructional steel);

(5) Sealant coating is suggested to apply on the contact surface;

(6) Direct contact between titanium part and the part made from lead, zinc, cadmium, silver, tin, bismuth, etc. is not allowed. For example, when titanium contacts a cadmium-plated part, it causes cadmium embrittlement, the susceptibility of which is much greater than high-strength steel. If the titanium part and cadmium-plated part have to be assembled together, the operating temperature should be lower than 100-150°C.

(7) In engineering application, galvanic corrosion of titanium is correlated with the coupled metal and the area ratio. Generally, the corrosion of the coupled metal is accelerated when the titanium/metal area ratio is higher than 4; and the corrosion is notably alleviated when the area ratio is lower than 1. Cast/carbon steel suffers serious galvanic corrosion when coupled with titanium, the corrosion rate rises by 9 times when the titanium/steel area ratio is 20, the corrosion rate reaches 1 mm/y in still seawater, and a few millimetres per year in flowing seawater.

2.7.4 Hydrogen Brittlement

Hydrogen brittlement of titanium alloys is closely associated with the materials processing and service environment. The main methods to prevent hydrogen brittlement are taken into account from the following aspects:

(1) Decrease of hydrogen content in titanium alloy and prevention of hydrogen absorption during processing

Smelting, machining, heat treatment and pickling must be carefully controlled to lower the hydrogen content in titanium alloy. The hydrogen content in titanium alloy ingot falls below 100 ppm after vacuum melting, but increases after subsequent processes, for example, electric furnace/flame furnace heating.

Protective coatings, such as a glass coating containing Ti or SiO_2-B_2O_3, water-soluble glass coating, high-temperature coating, etc., are applied on the titanium surface to prevent hydrogen absorption during heating at high temperatures. Due to their good performance in protection, water-soluble glass coating and high-temperature coating have been applied in industry. The atmosphere in the furnace should be carefully controlled, and generally a slightly oxidizing atmosphere is adopted.

Usually, welding is most likely to cause hydrogen absorption, especially on titanium parts with a complicated structure. Due to the improper design of the weld joint or lack of protective atmosphere, the weld seam is often polluted by hydrogen. Therefore, one should pay attention to: proper welded joint design; efficient gas shield and cooling on both sides of the titanium part; argon having sufficiently high purity; vacuum dehydrogenation of welding stick; careful cleaning of welding groove.

For some small size welding parts, integral annealing is one of the effective measures to reduce residual stress and remove hydrogen after welding. Under vacuum pressure (1.33×10^{-2} Pa), the hydrogen content of the weld seam and base metal is below 10 ppm.

(2) Reduction in hydrogen embrittlement susceptibility

Aluminum can be added into titanium alloys to increase hydrogen solubility, prevent the formation of titanium hydride, and then reduce hydrogen embrittlement susceptibility. Some near α titanium alloys, such as Ti-6Al-2Sn-4Zr, Ti-5Al-6Sn-2Zr-1Mo-0.25Si, have attracted much attention in recent years. In these alloys, β stabilizing elements, such as Mo and Si, are added to achieve a certain amount of β phase, which significantly increases the hydrogen solubility and reduces hydrogen embrittlement susceptibility.

Correct heat treatment is also very important. After cold machining and hot processing, vacuum annealing should be employed to release stress to form equiaxed grains, eliminate remnant β phase, and finally reduce hydrogen embrittlement susceptibility. Vacuum annealing can also be adopted to hydrogen absorption parts to get rid of hydrogen.

(3) Applicable environment

Titanium, like any other metal, has its own applicable environment. It was proposed that in dry or wet hydrogen atmospheres no hydrogen embrittlement of titanium occurs below 71°C due to slow diffusion of hydrogen; it is not recommended to be used at temperatures above 316°C due to its low strength. The practicability at temperatures between 71°C and 316°C is determined by the circumstances, and need to be tested before practical application. A certain amount of oxygen and water in the atmosphere could prevent hydrogen absorption and iron pollution on the surface. Thermal oxidation and anodic oxidation could prevent titanium from hydrogen absorption in hydrogen atmosphere.

In electrolyte solutions, consideration must be given to the conditions of hydrogen absorption, such as chemical reaction, uniform corrosion and local corrosion, etc. Generally, titanium is applied in oxidizing media, neutral media, weakly reducing media, or reducing acids containing oxidizing agent. Under these environments hydrogen absorption does not occur or occurs at a very slow speed. But if pollution, surface defect, local corrosion or unusual service conditions exist on the titanium surface, hydrogen may be evolved and be absorbed, resulting in embrittlement.

(4) Surface treatment

Presently it is hard to prevent hydrogen absorption of titanium in a highly corrosive medium only by surface treatment. But it is effective to improve hydrogen resistance, especially in neutral, weakly reducing media. The surface treatments to enhance the corrosion resistance of titanium include laser radiation, amorphization, alloying, noble metal coating, ion implantation, etc. In addition, chemical oxidation, thermal oxidation and anodizing are used to thicken and intensify the oxidation film.

(5) Addition of oxidizing agent

If the process conditions permit, an oxidizing agent, such as air, Fe^{3+}, Ni^{2+}, NO_3^- and O is added to the corrosive medium. They act as a corrosion inhibitor for titanium. Harmful ions, such as F^- should be removed from the medium.

(6) Other measures

Surface damage and iron pollution should be avoided during installation and maintenance of titanium parts. It should be prohibited to use steel tool to strike, fasten or descale titanium parts. Periodical acid pickling (with a corrosion inhibitor) should be applied. Correct structural design of titanium parts is necessary to eliminate crevices and stagnant regions, preventing galvanic corrosion. When cathodic protection is applied, the potential of titanium must be higher than the hydrogen evolution potential or a suitable sacrificed anode must be selected. Periodical inspection of titanium equipment is important.

2.7.5 Stress-corrosion Cracking

Finally, the following measures are used to reduce or eliminate SCC susceptibility of titanium alloys in a certain medium.

(1) Elimination of residual stress

Integral or local annealing is adopted to eliminate residual stress. But negative effects of heat treatment on material strength, plasticity and toughness must be considered.

(2) Alloying composition

Depending on different scenarios, the addition of the proper amount of Pd, Mo or Ru into titanium alloys may be considered to improve its SCC resistance.

Chapter 3 - Cavitation Erosion

3.1 General Understanding of Cavitation Erosion

Cavitation is the process of nucleation, growth and collapse of bubbles in a liquid or in an interface of solid-liquid when the local pressure is lower than the saturated vapour pressure. The high impact pressure, that generated by the implosion of bubbles, causes damage on the surface of solid material (Bardal 2004, Jiang et al 2003). This phenomenon is known as cavitation erosion (also called cavitation corrosion). It has been similarly defined as the localized attack that results from the collapse of voids or cavities due to turbulence in a liquid at a metal surface. It normally occurs in a fluid dynamic system, such as a hydraulic turbine, pump, valve, hydrofoil or ship propeller, etc.

Figure 3-1 illustrates the steps of a typical cavitation erosion process on the surface of metal.

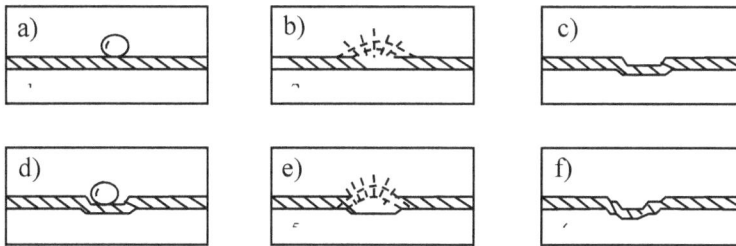

Figure 3-1 Steps of cavitation erosion process on the surface of passive metal

This process can be roughly divided into the following steps:
a) Vapour bubbles form near a metal surface;
b) The bubble collapse produces an intense impact against the metal. This possibly induces local plastic deformation and damage of the protective film at some spots of the metal surface;
c) The exposed metal surface suffers from corrosion, meanwhile the protective film re-forms on the surface;
d) New bubbles form near the damaged metal surface;
e) The bubble collapse causes the damage of re-formed protective film;
f) Re-exposed surface is attacked and new protective film grows again. Repeated impacts result in the cumulative damage.

Cavitation erosion is a complex phenomenon involving the joint interaction of mechanical, chemical and electrochemical factors. Currently there are several theories to explain the cavitation damage mechanism: mechanical effect, corrosion effect and thermal effect (Davim 2012).

1)   Mechanical effect

In the bubble collapse process, mechanical action is the most important factor causing cavitation damage. Two theories exist: shock wave and micro-jet. In the shock wave theory, bubble collapse produces pressures (shock waves) high enough to cause damage. In terms of the other theory, liquid micro-jets formed by imploding air bubbles are the main factor causing the damage. These physical phenomena lead to mechanical degradation of materials.

2)   Corrosion effect

In recent years, the study on the mechanism of cavitation damage shows that the role of chemical corrosion cannot be ignored. Even though the mechanical damage force may be lower than the strength required to destroy the material surface, the impact is enough to eliminate the protective film at some spots of the surface. This enables the metal surface to be exposed in corrosive medium and produces localized corrosion damage. The cavitation bubbles easily nucleate at these damaged sites and further accelerate the damage. The synergetic effect of mechanical action and corrosion can produce more damage than if each acts separately (Barik et al 2009, Kwok et al 2000).

3)   Thermal effect

It has been acknowledged that the mechanical damage process of cavitation erosion is often accompanied by thermal effect. The temperature of the bubbles at the collapsing moment reaches hundreds of degrees centigrade. This thermal effect may reduce the strength of the metal surface, causing more damage.

It is necessary to discuss the characteristics of cavitation erosion from various angles to get a better understanding of the mechanism and controlling parameters. This can provide information and data for the guidance of predicting and preventing cavitation erosion.

3.2 Methodologies of Cavitation Erosion Investigation

The cavitating flow cannot be measured properly because it is in association with the spatial and temporal random succession of pressure waves emitted during the collapse of bubbles. Thus, cavitation erosion degree of material is used as an indicator to evaluate the aggressiveness of the flow on the material. In addition, surface morphologies, chemical composition, electrochemical properties and surface mechanical properties are also investigated to explore the cavitation erosion characteristics and clarify the cavitation erosion mechanism.

3.2.1 Measurement of Cavitation Erosion Degree

There are a series of measurement methods (Lin et al 2016), such as mass loss, volume loss, pit depth and pit number, etc. Table 3-1 gives a comparison for different methods.

Table 3-1 Characterization methods of cavitation erosion degree

Method	Measured parameter	Advantage	Disadvantage
Mass loss	Mass difference before and after cavitation erosion (g/h)	Simple and easy for material with large mass loss	Not applicable for materials with large plasticity
Volume loss	Volume difference before and after cavitation erosion (cm³/h)	Simple and easy for material with large volume loss	Not applicable for materials with large plasticity and small volume loss
Pit depth	Average pit depth in a certain area (cm or μm)	Applicable for material with small mass loss	Difficult to accurately measure if the pit size is not identical
Pit number	Pit number per unit time and unit area	Applicable for material with small mass loss	The test duration cannot be too long

1) Mass loss

It can be calculated by the mass difference before and after the test. The mass loss per unit time is called cavitation erosion rate. The common unit is g/h. This method is relatively simple and easy (Brunatto et al 2012). It is applicable for higher mass loss materials. The error is large for materials with large plasticity.

2) Volume loss

It can be obtained by the volume difference before and after the test. The common unit is cm³/h. This method does not apply when the plasticity of the material is large and the material is only plastically deformed but has no mass loss or small volume loss after cavitation erosion.

3) Pit depth

The depth damaged after cavitation erosion is an important parameter to measure the degree of cavitation erosion. But because the depth varies with position on the surface, the size of each pit is not the same.

Accordingly, the average depth of cavitation erosion in a certain area is commonly used.

4) Pit number

The pit number per unit time and unit area is another parameter which can be used for representing the degree of cavitation erosion. It was proposed by Prof. Robert T. Knapp (1955). As long as the test duration is not too long and no pits overlap, the counting accuracy of the pit number is not affected.

5) Other methods

Some other parameters are considered to characterize and measure cavitation erosion degree, such as a localized elastic parameter H_e. According to the variation of this parameter, phase transformation and cavitation erosion resistance of material can be assessed (Wang and Zhu 2003). The relationship between mass loss and localized elastic parameter can be expressed by Eq.(3.1):

$$\Delta m = 3.4073 + \{968.8421/[4(He - 45.7451)^2 + 2.5778]\} \qquad (3.1)$$

where Δm is the cumulative mass loss in mg; H_e is the recoverable penetration depth in μm, characterizing the localized surface elasticity in two dimensions.

Based on Knapp's initial idea, cavitation pit counting techniques have been developed using optical method or laser profilometry with the aid of automatic analysis software (Patella et al 2000, Osterman et al 2009). The parameters, volume damage rates "V_d" and/or pit number rates "N_d" obtained from pit counting techniques can be related to cavitation erosion damage rate. Among these techniques, 3D methods are employed to give the pit geometries, such as pit depth, shape, and volume damage rates V_d. In 2D methods, some assumptions are considered: the pit is a segment of a sphere and the ratio between pit depth and pit radius is constant.

But these techniques are only applied for cavitation erosion analysis during incubation period. After a longer test, the influence of overlapping pits has to be considered. Hence some authors have begun to explore pit separation means to treat superposed impacts. Details have been described by Patella, R. F. et al (2000). A statistical analysis is also performed by others (Osterman et al 2009) to get the distribution of the number and size of pits, and consequently the distribution of cavitation erosion degree. More research is needed to minimize the influence of analysis parameters and some corrections are needed to obtain a reliable estimation of volume damage rates.

3.2.2 Structure Characterization of Cavitation Erosion

The process for cavitation damage on the surface of materials can be investigated in detail by various technologies, such as optical microscope (OM), scanning electron microscopy (SEM), atomic force microscopy (AFM), profilometer, confocal laser scanning microscopy (CLSM), energy dispersive spectrum (EDS), wavelength dispersive spectrometry (WDS), X-ray diffraction (XRD), and X-ray photoelectron spectroscopy (XPS). In the following, a few analyses are discussed using these technologies.

1) Morphology examination

2D attack morphologies on the surface or cross-sectional area in large and small scale can be observed by SEM. The damage features, including plastic deformation, pit growth, cracks, crystal grain size change, localized corrosion characteristics, microstructure, corrosion or wear damage traces, can be identified (Zheng et al 2008).

In general, OM is used to observe pit distribution, pit shape, microstructure, and corrosion or wear damage traces on the surface and cross-sectional appearance with a lower magnified factor.

2) Damage quantification

The following information: ① 2D and 3D images, ② surface roughness, ③ surface profile and depth measurement, can be provided by AFM. Normally, the maximum scan range is 110 μm×110 μm in the X and Y directions and 22 μm in the Z direction. It cannot be applied to observe larger features (Yong et al 2011). Besides AFM, profilometer can be employed to obtain the profiles of the eroded surface. According to the profile curves after cavitation erosion of different durations, the roughness values, Ra, are calculated. This is helpful to determine the different periods of cavitation erosion.

CLSM is an optical imaging technique for increasing optical resolution and contrast of a micrograph by means of adding a spatial pinhole placed at the confocal plane of the lens to eliminate out-of-focus light. 3D structures can be reconstructed from the obtained images by collecting sets of images at different depths within a thick object. This technique has gained popularity in measuring volume, surface area and depth of eroded pits on the electrode surface after polarization test in a cavitation erosion study (Fernández-Domene 2010, Fernández-Domene 2011). This can help determine the cavitation erosion degree from a microscopic point of view.

3) Composition analysis

Quantitative concentration of elements before and after cavitation erosion test can be examined by EDS and WDS. Element composition and distribution can be determined by point, line and area scanning.

The phases on the surface can be analyzed by XRD. The changes of microstructure characteristics during cavitation erosion can be investigated via XRD patterns.

XPS can be applied to a broad range of materials and provides valuable quantitative and chemical state information from the surface of the material being studied. It can give the information of microstructure changes or phase transformation and correlate material microstructure to cavitation erosion behavior.

3.2.3   Examination of Surface Mechanical Properties and Electrochemical Behavior

For some materials, hardness (macro and micro) is a good cavitation erosion resistance indicator. The hardness on the surface or the hardness profiles along the depth can be measured by using a nano tester or a Vickers microhardness tester to indicate the effect of impact on the wear resistance of material and work-hardening ability.

Corrosion may incur deterioration of mechanical properties of the surface layer. Local elasticity, surface hardness and work-hardening ability are the important controlling factors for cavitation erosion resistance of metallic material. The change of surface mechanical properties of the metallic material has an important relationship with its cavitation erosion behavior. The hardness represents resistance of a material to deformation, indentation, or penetration caused by scratch and impact. The elastic modulus is a measure of stiffness of a material when an external force is applied to it. The ratio of nano-hardness to elastic modulus of corrosion layer on the metal surface is a dimensionless parameter, which can be used to comprehensively describe the mechanical property of surface layer during cavitation erosion (Li et al 2012).

Electrochemical techniques, such as open circuit potential (OCP) monitoring, current monitoring at different applied potential, potentiodynamic polarization measurement, and cyclic potentiodynamic polarization are useful tools to evaluate the corrosion performance and repassivation tendency of passive materials during cavitation erosion (Zheng et al 2007). The following parameters from these electrochemical curves can be obtained: corrosion potential (E_{corr}), corrosion current density (i_{corr}), pitting potential (E_p), passive current density (i_p), repassivation current density (i_{rp}), repassivation potential (E_{rp}), hysteresis anodic loop (E_p-E_{rp}), the extent of the passive region.

The results from surface mechanical properties and electrochemical measurements can be combined to discuss the interaction mechanism between the deterioration of mechanical properties and corrosion failure.

3.3 Factors Affecting Cavitation Erosion

As mentioned earlier, the cavitation phenomenon is complicated, and there are many factors affecting the cavitation erosion of metallic materials. They comprise of material performance and the external environment (Lin et al 2016).

3.3.1 Material

3.3.1.1 Mechanical Properties

Generally, a metal with high hardness and hardenability has a relatively high cavitation erosion resistance. The experiments of cavitation erosion of CP-Ti and Ti-6Al-4V alloy in 3.5% NaCl solution showed that the cavitation erosion resistance of Ti-6Al-4V with higher hardness was better than that of pure titanium (Mochizuki 2007, Guan 2010). The studies of cavitation erosion of Cr-Ni-Mo and Cr-Ni-Co alloys in artificial seawater gave similar results. But this is not always true. Some materials with low hardness also show good cavitation erosion resistance. This is ascribed to their good work-hardening property. For example, Cr-Mn-N stainless steel with low hardness in distilled water was more resistant to cavitation erosion compared to 0Cr13Ni5Mo stainless steel which had high hardness (Zheng et al 2008); the cavitation erosion resistance of tin brass with low hardness was significantly superior to that of aluminum bronze in 3.5% NaCl solution. Accordingly, both hardness and work-hardening properties are the main factors affecting the cavitation erosion resistance of materials.

In addition, the cavitation erosion resistance of materials is relevant to the yield and the tensile strength. Low yield and tensile strength give rise to larger cumulative mass loss, and vice versa. High impact work, elongation and elasticity of materials are beneficial to the cavitation erosion resistance, such as 00Cr13Ni5Mo martensitic stainless steel, cast titanium alloy, and TiNiNb alloy. Increasing the relative content of lower bainite in steel can improve the cavitation erosion resistance, mainly because of its high toughness and effective impact resistance in preventing the propagation of pits and cracks. Researchers (Liu et al 2012) also found that in contrast to the traditional 304 austenitic stainless steel, a new-type, Cr-Co-Ni-Mn austenitic stainless steel, with low plasticity, high yield and tensile strength had a longer incubation period and better cavitation erosion resistance.

3.3.1.2 Microstructure and Chemical Composition

The chemical composition of material usually play an important role in cavitation erosion resistance. The addition of 5%, 7%, and 10% (mass fraction) Ni and Mn into Fe-12Cr-0.4C alloy could produce different effects on cavitation erosion resistance (Park et al 2013). The cavitation erosion resistance is improved with a decrease of Ni content and a increase of Mn content. Moreover, the alloy with addition of Mn is more resistant to cavitation erosion than that with addition of Ni. Austenitic stainless steel containing higher Co and Cr elements has low stacking fault energy. Meanwhile the strain generated in the cavitation erosion process causes the transformation from austenite to martensite, which absorbs the impact energy induced by the bubble collapse. These result in the prolonged incubation period and high cavitation erosion resistance of Cr-Co-Ni-Mn austenitic stainless steel.

The crystal structure and grain size are also important influential factors for cavitation erosion resistance (Bregliozzi et al 2005). The fcc metal is prone to plastic deformation under the pressures produced by cavitation shock waves or micro jets, and is not sensitive to high strain rate, which leads to a long incubation period and good cavitation erosion resistance. The bcc and hcp metals are very sensitive to strain rate, and their cavitation failure mode is generally fast transgranular fracture and cleavage fracture. They are not resistant to cavitation erosion. However, if the martensitic transformation for hcp metal occurs during cavitation erosion, the resistance is greatly improved. On one hand, the martensitic transformation consumes a large amount of energy produced by a bubble collapse, and on the other hand the strength and hardness of the martensitic phase is relatively high, enhancing the work-hardening ability of the material. In addition, the finer the grain is, the better the cavitation erosion resistance is. The grain refinement may bring about the following benefits, which contribute to the enhancement of cavitation erosion resistance.

- Improvement of the material mechanical properties, including the strength, hardness and plasticity, to a certain extent;
- More uniform deformation and reduced crack formation and propagation as a result of local stress concentration; and
- Strengthening of grain boundary and work-hardening ability.

The cavitation erosion behavior and resistance differ from microstructure. For instance, the order of cavitation erosion resistance of phases in stainless steel is: austenite>martensite>ferrite. The damage of dual phase materials is mainly determined by a low strength phase (Liu and Chen 2009). It absorbs cavitation impact energy and mitigates the damage degree of a high strength phase.

3.3.1.3 Surface Topography

The surface condition of material affects not only the degree of cavitation erosion, but also the process of cavitation erosion.

It is generally believed that rough surfaces promote the occurrence of cavitation erosion (Chiu et al 2005). The rough surface contains micro structures, such as grooves, convex edges and concave bottoms, which accommodate more cavitation nucleus, resulting in the increase of the degree of cavitation. A large number of cracks and defects on the rough surface raise the probability of damage, and simultaneously convex edges are vulnerable to be broken and torn through strong impact. This ultimately leads to more severe damage on the rough surface than that on the smooth surface. However, regular undulate to some extent of the surface can slow down the effect of the micro jets.

3.3.2 External Environments

Some investigations indicated temperature and pH in environment have an influence on cavitation erosion behavior. For pure titanium and Ti-6Al-4V alloy, the volume loss rate increased with rising temperature in seawater when the temperature was in the range of 16-43°C. The studies of effect of temperature and pH on cavitation erosion of super duplex stainless steel in 3.5% NaCl solution suggested that the average erosion depth rate increased with an increase in temperature, peaked at 50°C, after which a decrease was observed. With regards to pH, the average erosion depth rate declined as pH increased, at a pH of 9 it was lowest, and then it went up with the increase of pH (Kwok et al 1997).

Other external influential factors include physical properties, corrosivity and flow rate of the medium. The cavitation erosion damage is more serious in a medium with larger surface tension, lower viscosity and compressibility. A greater flow rate of liquid can produce a larger pitting corrosion rate in the incubation period. The mass loss of 20SiMn low alloy steel after a cavitation erosion test of 2.5 h in 3.5% NaCl solution was about two times the value of that in distilled water.

3.3.3 Summary

Table 3-2 summarizes the principal effects of internal and external factors.

Table 3-2 Major factors affecting cavitation erosion resistance

Factors	Indicators	Principal effects
Mechanical properties of materials	Microhardness	Generally, the higher microhardness, the more resistant to cavitation erosion
	Work hardening property	The higher work-hardening property, the more resistant to cavitation erosion.
	Yield/tensile strength	The higher yield/tensile strength, the more resistant to cavitation erosion.
	Toughness	The higher toughness, the more resistant to cavitation erosion.
Chemical composition and microstructure of materials	Chemical composition	Mn, Co and Cr with higher content and Ni with lower content would increase the resistance of cavitation erosion.
	Crystal structure	The material with fcc structure or containing austenitic phase is more resistant to cavitation erosion; the martensitic transformation during cavitation erosion process can improve the resistance; for multi-phase material, the resistance is mainly determined by the lower intensity phase.
	Crystal size	The material with fine grains has good mechanical properties, which is beneficial to the cavitation erosion resistance.
	Stacking fault	The lower stacking fault energy, the more resistant to cavitation corrosion
Surface topography	Surface roughness	Normally, the material with rough surface is more prone to cavitation erosion. However regularly undulating surface to some extent may prolong incubation period, improving cavitation corrosion resistance.
External environments	Medium flow rate	The larger flow rate promotes cavitation erosion.
	Medium temperature	With an increase in temperature, the average erosion depth rate increases first, then reaches the peak value, after that decreases.
	Medium corrosivity	The concentration and pH of corrosive medium significantly affect the degree of cavitation erosion.

3.4 Cavitation Erosion Behavior of Titanium Alloys

3.4.1 Background

With the acceleration of industrialization in the world, water pollution is becoming more and more serious, and the problem of solving the shortage of fresh water resources is increasingly being put on the agenda to address by the governments of many countries. 97% of the water on earth is salty, so using seawater would become a solution to the lack of fresh water. However, seawater has an average salinity of 3.5%, which is too high for drinking, industry or agriculture. Desalination is the process of removing the salt from seawater to turn the sea water into fresh water. This technology has gained more and more attention in the world. There are many methods of seawater desalination, and one of them is multi-effect distillation. Aqueous lithium bromide (LiBr) is an environmentally-friendly absorbent. Since the concentrated LiBr solution possesses strong water absorption characteristics, it absorbs water to become a dilute solution, at the same time, heat is released which can provide a source of heat and power for multi-effect distillation seawater desalination technology (Mandani et al 2000). Therefore, the application of environmentally friendly heavy LiBr solution in seawater desalination is a new design concept.

However, aqueous LiBr is a strong corrosive liquid. The cavitation phenomenon is easy to occur at high flow velocity and at pressures below the saturated vapour pressure of this liquid. Eventually, it may lead to cavitation erosion of the metallic components in bends and narrows of LiBr refrigeration absorption machine.

Cavitation erosion damage produces the problems of reduction in operation efficiency, service life and reliability. Therefore, more corrosion resistance metallic materials are required in the construction of LiBr absorption machines (Figure 3-2). Titanium alloys as passive metals exhibit high corrosion resistance in a wide range of aggressive media, such as halide solutions, which are expected to become the making materials of some components in LiBr refrigeration absorption machine. But the protective passive film on the surface may be removed under the action of mechanical force, causing the decrease of corrosion and wear resistances, therefore whether titanium materials can be used in LiBr absorption machines or not depends on their resistance to cavitation erosion.

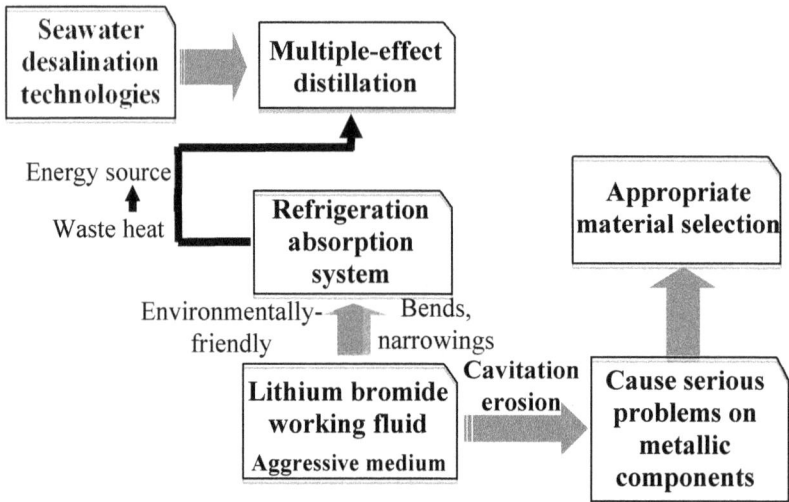

Figure 3-2 Cavitation erosion in LiBr refrigeration absorption machine

The corrosion behavior of titanium alloys has been widely studied and there is extensive literature related to corrosion of titanium alloys in various corrosive environments and media (see details in Chapter 2). However, much less attention has been paid to their resistance to tribo-corrosion where there is a conjuction of corrosion and mechanical degradation (e.g. abrasion, erosion, cavitation, etc.) (Neville and McDougall 2001). Thus, it is crucial to investigate their cavitation erosion behavior in the working fluid of aqueous lithium bromide solution. The recent reseach on the cavitation erosion behavior of Ti-6Al-4V and CP-Ti conducted by the research group led by Dr. Lin (the first author of this book) is presented in section 3.4.2. to section 3.4.4.

3.4.2 Laboratory Testing

3.4.2.1 Ultrasonic Vibration Apparatus

In the early stages, research on the cavitation erosion behavior of materials was conducted using in-situ test equipment in the field. However, damage may occur after a long duration of exposure. Thus, this kind of test method is uneconomical and inefficient.

Experimental studies in the laboratory offer a convenient means of evaluating the cavitation erosion performance within a short period of time. Such accelerated test equipment includes ultrasonic vibration, rotating discs, liquid jet, and reciprocating piston, etc. The ultrasonic vibration is the simplest and the most frequently used laboratory technique for testing cavitation erosion characteristics of materials and was standardized by the American Society for Testing and Materials (ASTM G32-16 2016). In this standard, the ultrasonic vibration setup is described and the relevant parameters are provided.

The modified ultrasonic cavitation erosion test apparatus is shown in Figure 3-3. In the picture, the left part (1) is the ultrasonic generator and the right part (3) is the temperature control device. The inside configuration and the schematic illustration of the middle part (2) is depicted in Figure 3-4 and 3-5, respectively. This ultrasonic vibration apparatus has a generator, a transducer, a horn, a beaker to hold the test liquid and a temperature controller bath. The generator is coupled to the transducer which converts the electrical signal into a mechanical vibration. This signal is magnified through a horn down to the tip. This transmission allows the horn to vibrate at high frequency, generating an excitation in the liquid leading to a bubble generation in the clearance. Inside the beaker, a specimen holder designed to prevent a specimen from moving in the process of cavitation erosion is made up of PTFE material. The specimen is fixed by three fasteners. Additionally, three props are adjusted to ensure that the whole surface of the specimen is kept level. The illustration of the specimen holder is shown in Figure 3-6. Experimental arrangement of the cavitation erosion test is further modified to combine the electrochemical test system and the electrochemical cell in order to perform electrochemical monitoring during the cavitation test.

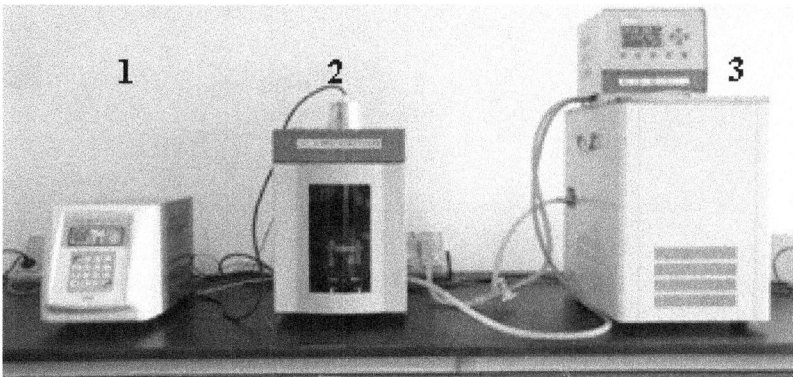

Figure 3-3 Picture of ultrasonic cavitation erosion test apparatus

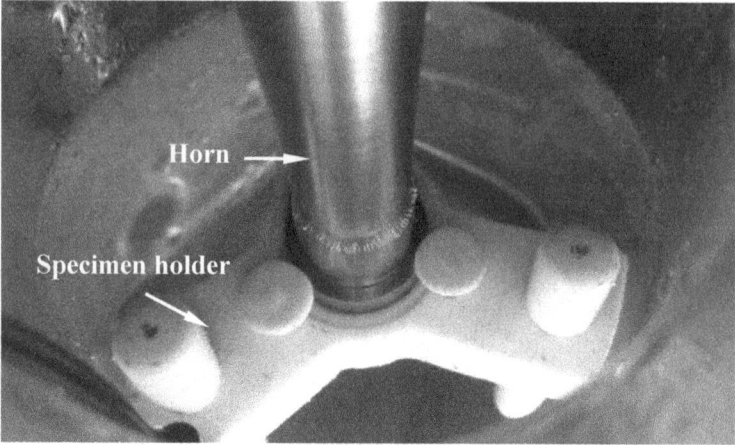

Figure 3-4 Configuration inside part 2

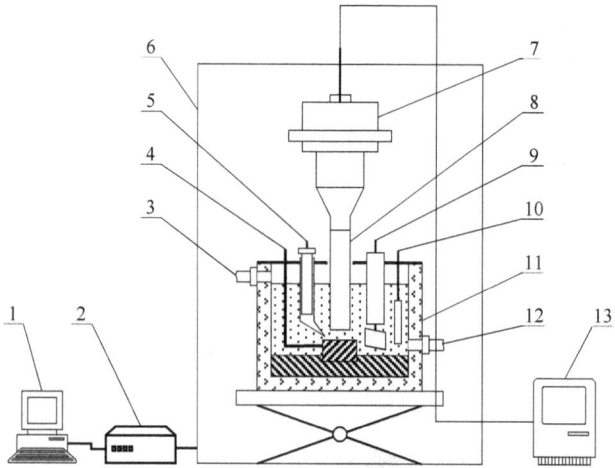

1-computer; 2-electrochemical test system; 3-water outlet; 4-working electrode; 5-reference electrode; 6-sound proof enclosure; 7-transducer; 8-horn; 9-auxiliary electrode; 10-temperature sensor; 11-cooling bath; 12-water inlet; 13-ultrasonic generator

Figure 3-5 Schematic of part 2

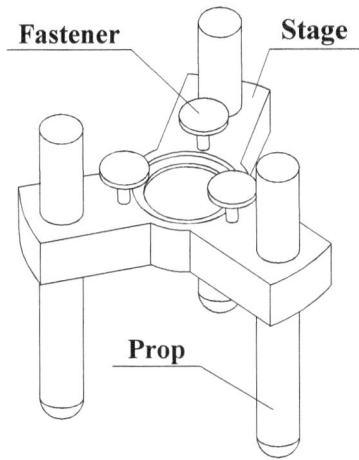

Figure 3-6 Schematic of specimen holder

3.4.2.2 Cavitation Erosion Test

All cavitation erosion tests in this section were performed on the above apparatus. The transducer operated at 20 kHz with a peak to peak amplitude of 50 μm. The tip and specimen were immersed to a depth of 12±4 mm. The cavitation erosion tests were conducted in ditilled water and 55% LiBr solution, which was prepared by LiBr with 99% purity and pure water. The solution volume in the beaker was maintained at 300 mL. The temperature was controlled at 25°C with an accuracy of ± 2°C. The technical parameters for cavitation erosion test are listed in Table 3-3.

Table 3-3 Technical parameters for cavitation erosion test

Output power	500 W
Frequency of the ultrasonic transducer	20 kHz
Peak-to-peak amplitude	50 μm
Horn tip diameter	15.9 mm
Depth of the specimen immersion	12 mm
Distance between the face of the horn tip and the test specimen	0.5 mm

The material used was annealed Ti-6Al-4V and CP-Ti rods. Ti-6Al-4V had a chemical composition of 6.4% α-stable element aluminum, 4.1% β-stable element vanadium, 0.04% iron, 0.01% carbon, and the remainder titanium. CP-Ti was an unalloyed titanium with 0.14%Fe, 0.01%C. The specimen was Φ20 mm×5 mm in size. It was situated at a distance of 0.5 mm to the opposite vibrating horn tip with 15.9 mm in diameter.

3.4.2.3 Characterization Methods

The characterization methods included mass loss and surface roughness measurement, 2D and 3D morphological features examination, surface mechanical properties and electrochemical behavior analysis.

The mass of the eroded specimen was measured at 30-minute or 60-minute intervals. The mass loss rate is calculated by Eq.(3.2).

$$\upsilon = \frac{m_1 - m_2}{t_1 - t_2} = \frac{\Delta m}{t_1 - t_2} \tag{3.2}$$

where Δm is the mass loss in mg; m_1 is the mass after a testing duration of t_1 in mg; m_2 is the mass after a testing duration of t_2 in mg; t_1 and t_2 are testing duration in minute (min); υ is the mass loss rate in mg/min.

The surface roughness Rq and the profiles of eroded surface were measured by JB-6CA roughness profiler with a capable resolution of 0.001 μm and a contact pin of 2 μm radius (Figure 3-7). The mean depth of cavitation erosion was obtained from the profile curve.

A FEI Quanta 200 Scanning Electron Microscope (SEM) and an Agilent 500 Atomic Force Microscope (AFM) were used to examine the morphologies. Energy Dispersive Spectropmetry (EDS) coupled to SEM was applied to determine element composition and distribution on the surface. Hirox KH-7700 digital microscope was further used to examine 3D morphology characteristics. The digital camera, light source, LCD monitor, computer and software are all integrated into this system, which has solved the limitation of the small field of vision at high magnification in traditional optical microscopes (Figure 3-8). It has the advantages of enlarged observation (0-7000×) and high-precision resolution (0.35 μm). 3D structure and data can be reconstructed real-time.

A 401MVD Vicker microhardness tester (Figure 3-9) and a Stress 3000 X-ray stress analyzer with Ti-Kα radiation (Figure 3-10) was employed to measure hardness and residual stress, respectively. In each hardness measurement, an indenter was forced into the surface with a load of 100 g for 15 s.

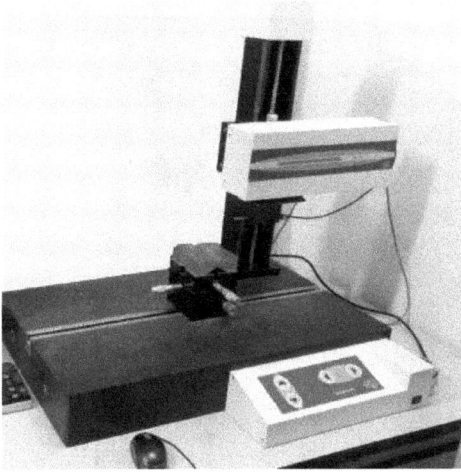

Figure 3-7 JB-6CA roughness profiler in Corrosion Research Laboratory at Nanchang Hangkong University

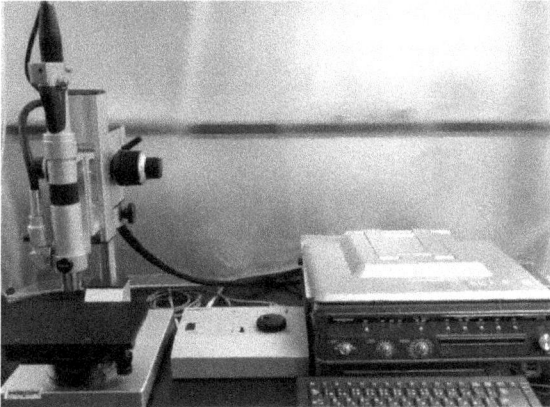

Figure 3-8 Hirox KH-7700 digital microscope in Corrosion Research Laboratory at Nanchang Hangkong University

The electrochemical testing was carried out using a AutolabPGSTAT30 electrochemical workstation. The selected electrochemical testing methods contained open circuit potential monitoring, potentiodynamic polarization and electrochemical impedance spectrum.

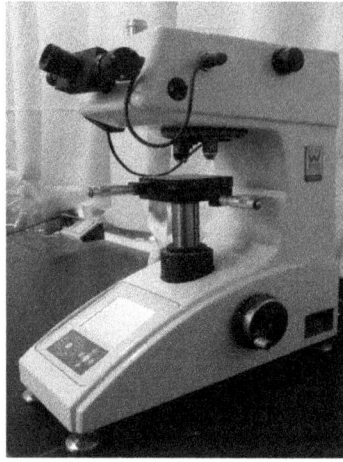

Figure 3-9 401MVD Vicker microhardness tester in Corrosion Research Laboratory at Nanchang Hangkong University

Figure 3-10 Stress 3000 X-ray stress analyzer in Corrosion Research Laboratory at Nanchang Hangkkong University

3.4.3 Cavitation Erosion Behavior of Ti-6Al-4V

The research results on cavitation erosion behavior of Ti-6Al-4V in aqueous LiBr solution is presented below (Zhang et al 2015a, Zhang 2015b, Zhao 2016). Based on these results, the cavitation erosion process of Ti-6Al-4V, the effect of microstructure on the cavitation erosion behavior and the combined effect of mechanical and electrochemical damage are discussed.

3.4.3.1 Total Mass Loss and Mass Loss Rate

The simplest way to evaluate the cavitation erosion degree is to obtain the mass loss during the cavitation eroiosn process.

Figure 3-11 and 3-12 show the cumulative mass loss-time curves and mass loss rate-time curves, respectively. In the initial 180 min of the test, the total mass loss is relatively small, less than 1.25 mg both in distilled water and LiBr solution. The total mass loss in LiBr solution is slightly greater than that in distilled water. After 180 min, a rapid increase of the total mass loss occurs. After 480 min of the continuous cavitation test, the total mass loss in distilled water is 5.38 mg, which is about 5 times the loss at 180 min. In LiBr solution, the mass loss also increases significantly, to 6.86 mg. The difference of the total mass loss between these two media is augmented. This indicates the contribution of corrosion in the cavitation erosion damage.

It can be clearly seen from the variation of mass loss rate during the 480 min test that there are three distinguishable stages: steady increase (0-180 min), rapid increase (180-300 min) and little change (after 300 min). These represent respectively the incubation period, acceleration period and steady-state period in the process of cavitation erosion of Ti-6Al-4V.

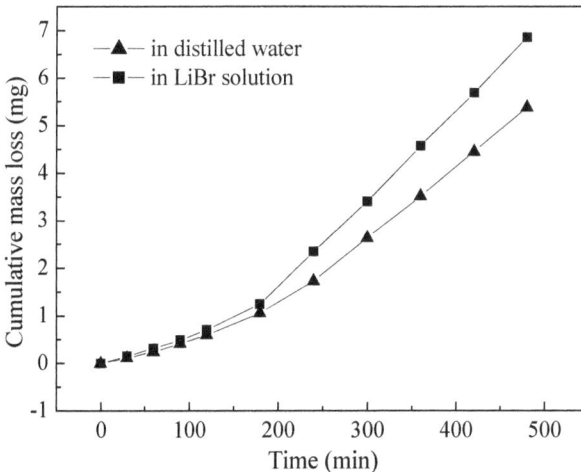

Figure 3-11 Cumulative mass loss of Ti-6Al-4V as a function of time

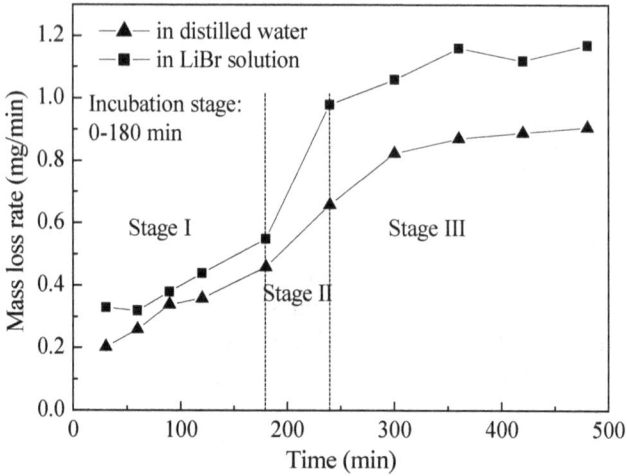

Figure 3-12 Mass loss rate of Ti-6Al-4V as a function of time

The mass loss rates in distilled water and LiBr solution exhibit a similar trend as time progresses. But the difference is that the mass loss rate in LiBr solution is higher than that in distilled water, especially in the stage of acceleration. From these results, it can be inferred that the mechanical effect is still a dominant factor of damage loss under cavitation condition, but the synergetic effect of cavitation and corrosion can promote the damage loss.

3.4.3.2 Surface Roughness and Mean Depth of Erosion

Recently, surface roughness of the eroded surface has been employed to analyze the process of cavitation erosion.

The evolution of surface roughness and the mean depth of erosion are illustrated in Figure 3-13. A comparison between Figures 3-11 and 3-13 shows that the variation tendency of mass loss rate and that of surface roughness versus erosion time is similar. The roughness-time curve and the mean depth of erosion-time curve in Figure 3-13 can also be divided into three stages.

In the initial stage (stage I), the roughness increases linearly. The surface roughness before the cavitation test is 0.221 μm. After 160 min of the cavitation test, it is increased to 0.462 μm. The mean depth of erosion is almost zero. The second stage is the transition stage (stage II). The increase rate of surface roughness starts to decline, while the mean depth of erosion goes up steadily. After 320 min of the cavitation test, the surface roughness is 0.559 μm and the mean depth of erosion is 0.088 μm. The following is the steady-state stage (stage III). The surface roughness has slight fluctuation, whereas the mean depth of erosion begins to have a pronounced increase.

It can be found that the three stages approximately coincide with the incubation, acceleration and maximum erosion rate stages in the mass loss rate curve.

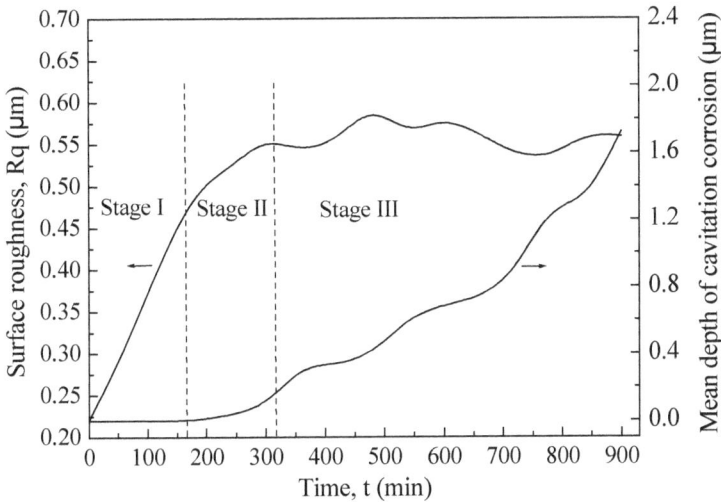

Figure 3-13 Variation of surface roughness Rq and mean depth of erosion for Ti-6Al-4V

3D surface profile of the eroded surface can provide microscopic visual information in concave-convex evolution on the surface during the cavitation erosion process. The results are demonstrated in Figure 3-14. The variation of concave-convex degree of eroded surface roughly corresponds to the three stages of the surface roughness-time curve. The depression areas in the 3D profiles (darker color in Figure 3-14) are pits induced by plastic deformation. Some small pits occur during early cavitation. As cavitation continues, the pits increase and grow deeper. Some pits gather in clusters. This results in the increase of concave-convex degree on the surface. During further cavitation, no obvious change in concave-convex degree arises.

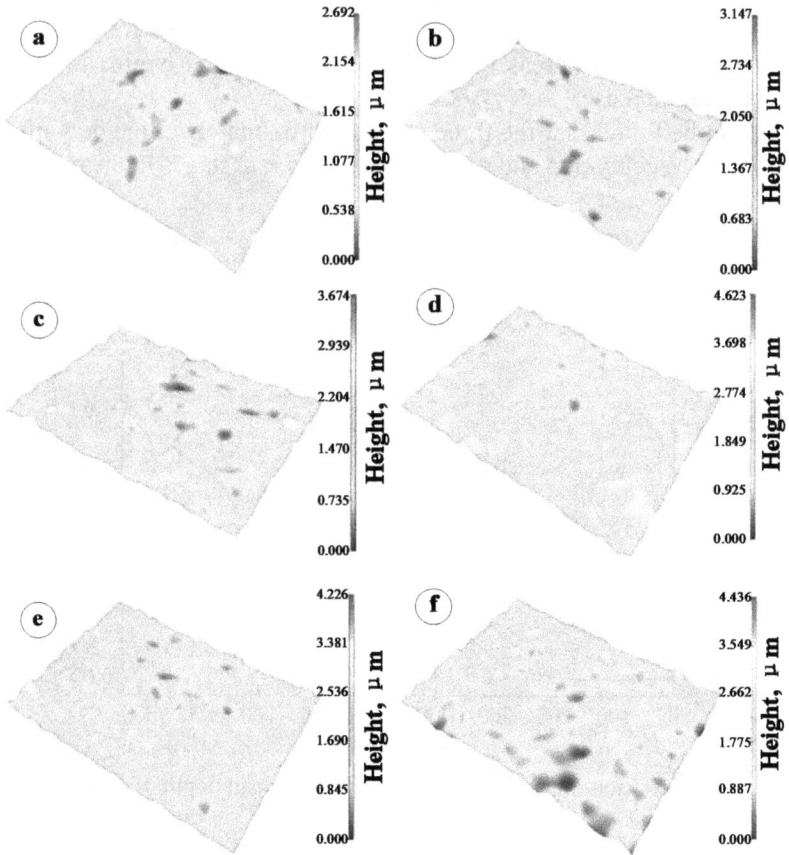

Figure 3-14 3D profile of cavitation erosion for Ti-6Al-4V at different test time: (a) 0 min, (b) 60 min, (c) 180 min, (d) 300 min, (e) 480 min, (f) 720 min

3.4.3.3 Characterization of Cavitation Damage

1) Surface and cross-section morphologies

After cavitation, the surface of Ti-6Al-4V loses metallic lustre and exhibits a gray black and rougher appearance.

The damage on the surface of Ti-6Al-4V in distilled water and LiBr solution observed from microscopic perspective is shown in Figure 3-15 (a and b), respectively. After 60 min of test, slight plastic deformation and shallow pits are generated. By contrast, the damage features in LiBr solution are more noticeable, as shown in Figure 3-15 (a-1) and (b-1).

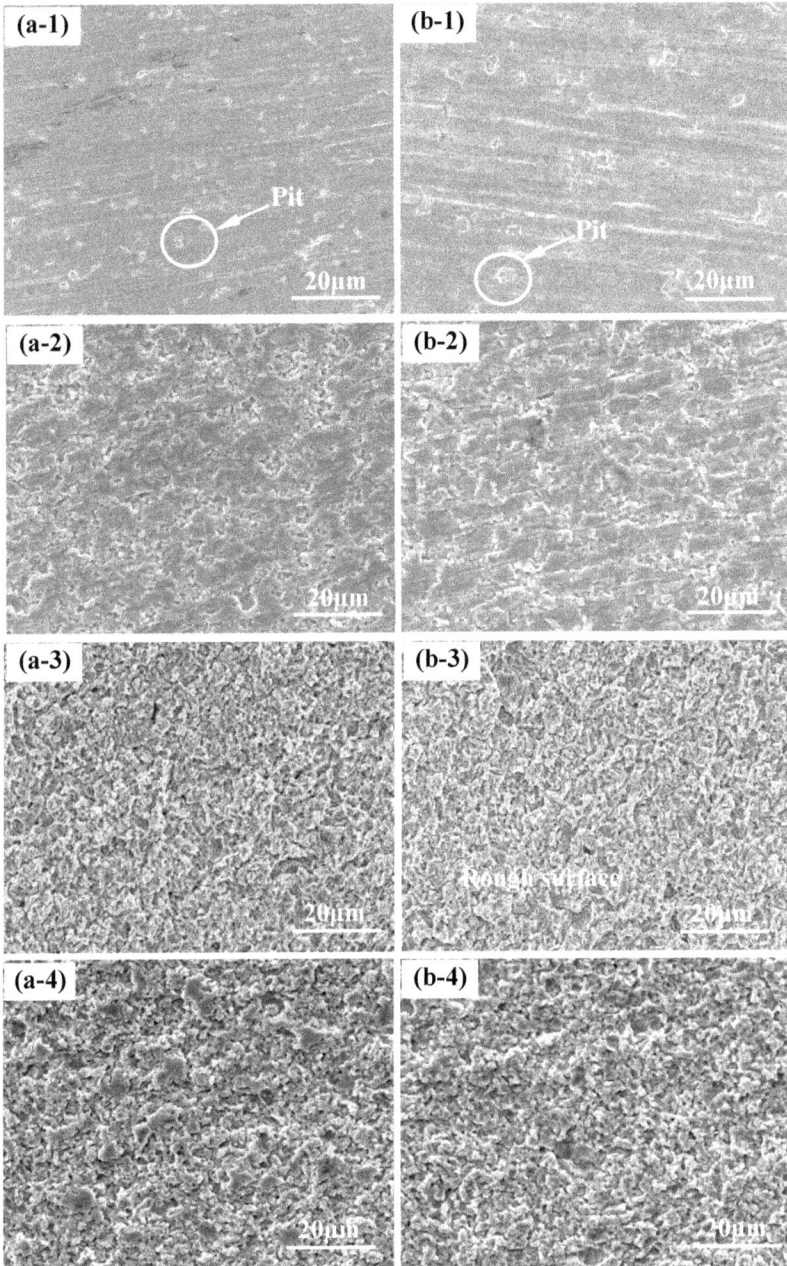

Figure 3-15 Morphologies of Ti-6Al-4V after cavitation erosion test in distilled water (a) and LiBr solution (b) for different time: (1) 60 min; (2) 180 min; (3) 300 min; (4) 480 min

After 180 min, the obvious plastic deformation and cracks take place. There is increasing amounts of pits, which spread to form clusters. This is accompanied by spalling of damaged surface materials in small quantities. This phenomenon is in accordance with the variation of mass loss rate. In LiBr solution, the pits are more concentrated and undulate folds appear, as seen in Figure 3-15(b-2). This phenomenon can be explained as follows. The forces from micro jets and shock waves produced by cavitation bubble collapses commonly concentrate to exert at the pits and cracks, leading to the degradation of mechanical properties at these areas and preferential dissolution in corrosive medium. The plastic deformation at the bottom between the protrusions can push the material to the edge to create folds. According to the mass loss rate curve, the maximum value is at 300 min of cavitation erosion test. From Figure 3-15 (a-3) and (b-3), the pits continuously propagate. The large area materials on the surface are damaged and fall off. With the test duration of 480 min, the whole surface falls off and presents a morphology of honeycomb.

The cross-section morphologies after 60 min, 180 min and 300 min of test are shown in Figure 3-16. After 60 min, a very thin layer of plastic deformation appears on the surface (Figure 3-16a). Figure 3-16b is the cross-section morphology in the end of the incubation period. The thickening of the plastic deformation layer is in companion with the propagation of cracks. Falling of material happens on small parts of the surface. After 300 min, the damage of the surface is more serious, substantial falling occurs on the whole surface, as shown in Figure 3-16c.

2) Effect of microstructures

The specimen was etched before cavitation erosion test in order to analyze the influence of phases in Ti-6Al-4V during the cavitation erosion process. The surface morphologies at high magnification were observed using SEM (Figure 3-17).

Ti-6Al-4V is composed of α and β phase. α phase is rich in aluminum (Al), while β phase mainly contains vanadium (V). Preferential attack occurs at α phase (hcp structure), where the strength is lower than that of β phase (bcc structure). Thus the passive film on α phase is easily broken to form pits under the mechanical action of cavitation. The fresh substrate on α phase is exposed in LiBr solution. α phase without passive film and β phase with passive film constitute the electrochemical corrosion micro cell on the surface. β phase acts as a cathode to promote the dissolution of α phase. This is the synergetic effect of mechanical action and corrosion.

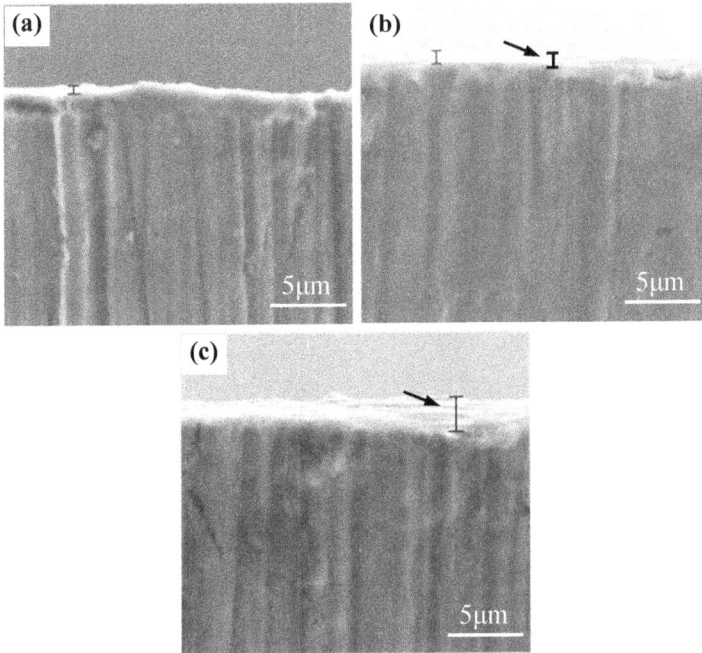

Figure 3-16 Cross-section morphologies of Ti-6Al-4V after cavitation erosion test in LiBr solution for different time: (a) 60 min; (b) 180 min; (c) 300 min

However, the area of α phase is greater than that of β phase, which avoids the phenomenon of small anode and large cathode. So, this can mitigate the promoting effect of corrosion to some degree. The preferential damage of α phase weakens the strength of the phase boundary between α phase and β phase. Some cracks and pits can be seen on the boundary after 60 min of the cavitation test. Furthermore, part of β phases falls off from the surface, which causes the removal of α phase (Figure 3-17c). After 180 min, there are apparent removal of α phase and β phase and many cracks on the eroded surface.

3) Elemental composition analysis

The EDS results of elemental composition on the surface at different time in distilled water and LiBr solution are given in Table 3-4 and 3-5.

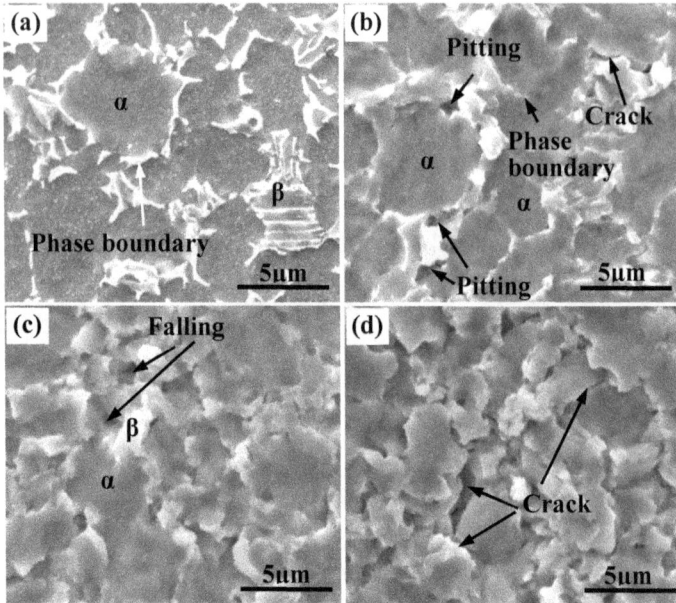

Figure 3-17 Morphologies of cavitation erosion of Ti-6Al-4V after etched in LiBr solution for different time: (a) 0 min; (b) 60 min; (c) 120 min; (d) 180 min

Table 3-4 Content of the elements on the surface of Ti-6Al-4V tested in distilled water (wt.%)

Test time (min)	Ti	Al	V
0	92.00	4.05	3.95
60	91.87	4.13	4.00
180	91.80	4.17	4.03
300	92.10	4.02	3.88
480	92.15	3.98	3.87

Table 3-5 Content of the elements on the surface of Ti-6Al-4V tested in LiBr solution (wt.%)

Test time (min)	Ti	Al	V
0	92.00	4.05	3.95
60	91.28	3.85	4.87
180	92.16	3.52	4.32
300	92.35	3.11	4.54
480	93.27	2.46	4.27

In distilled water, there is little change of the content of Ti, Al, V elements. In LiBr solution, the content of Al is decreased by 39.25% after 480 min. The variation of Al content against test time is plotted in Figure 3-18.

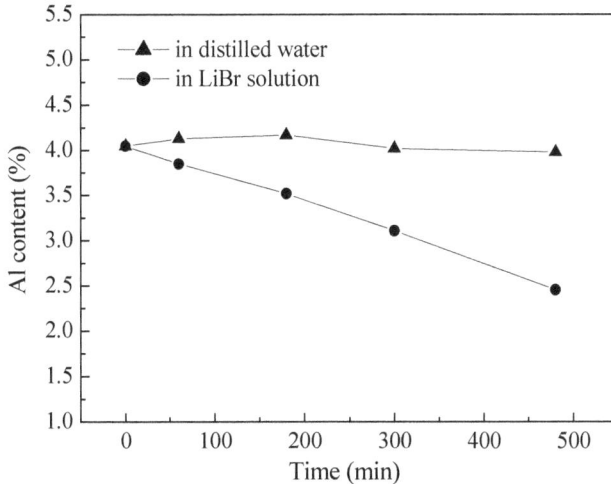

Figure 3-18 Variation of content of Al element under cavitation on the surface of Ti-6Al-4V tested in distilled water and LiBr solution

In distilled water, no corrosive medium exists. The damage on the surface majorly results from mechanical force. α phase and β phase are alternately removed. The damage degree of the two phases is basically the same.

In LiBr solution, the cause of the reduction of Al content after cavitation erosion is believed to be as follows.

- During the process of cavitation erosion in the presence of corrosive medium, Al element in α phase has more active potential and is easily corroded in the electrochemical reaction.
- Al element is used as a stable element of α phase. As mentioned above, α phase possesses a lower strength, which is first attacked. The combined action of mechanical effect and electrochemical corrosion can produce more damage on α phase.

Consequently, under the impact of repeated cavitation, more Al element is removed from the surface of Ti-6Al-4V.

3.4.3.4 Mechanical Properties of Eroded Surface

1) Microhardness

Surface hardness as an important measure for surface mechanical properties of material can reflect the capability to resist external impinging of material surface, so in general it is an important parameter to represent the cavitation erosion resistance of material. In addition, the change in surface hardness before and after cavitation erosion is also an indicator to measure the ability of absorbing cavitation impinging energy.

The microhardness of Ti-6Al-4V substrate is HV 330.7 MPa. The variation in the microhardness after different cavitation erosion duration is depicted in Figure 3-19. During the incubation period, the microhardness continuously increases and reaches a peak at 180 min. Based on the cumulative mass loss curve and the morphologies of the eroded surface, plastic deformation and work-hardening mainly occur when the surface is subjected to the action of shock waves and micro jets in the incubation period. There is less mass loss. At the end of the incubation period (180 min), plastic deformation is very serious, resulting in the growth of cracks and pits in a large area, so that the degree of work-hardening achieves the highest value on the surface, nearly 420 MPa. As the cavitation erosion enters the acceleration period, the large area materials fall off, which is accompanied by work-hardening on the sub-surface. The impact energy is not absorbed by the plastic deformation any longer, which makes the surface hardness decrease.

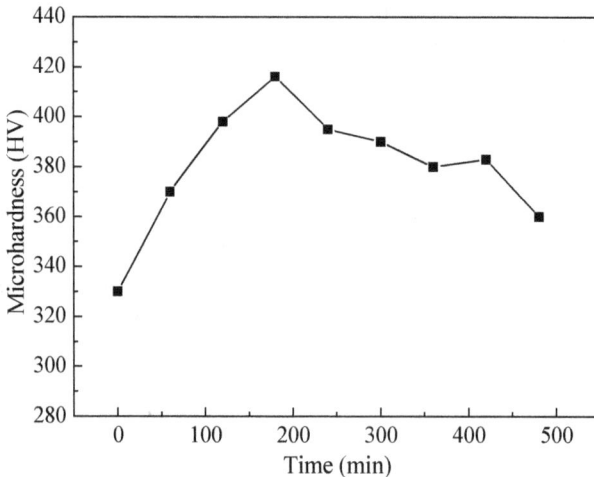

Figure 3-19 Microhardness on the surface of Ti-6Al-4V after different cavitation erosion durations in LiBr solution

The measurement of microhardness in the cross section was carried out to indicate the work-hardening of material after cavitation erosion. The measurement was taken at equally spaced points from the eroded surface to the inside of the substrate. Figure 3-20 is an optical image showing microhardness indentation marks from the surface to the substrate.

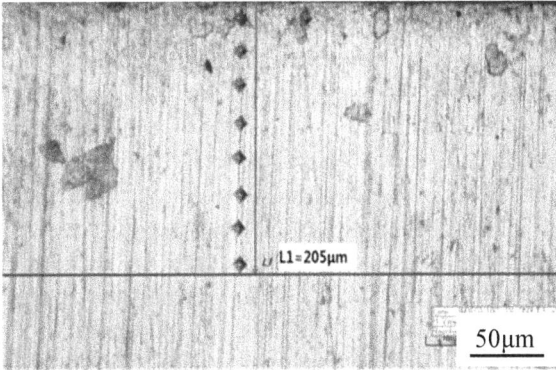

Figure 3-20 Optical image showing microhardness indentation marks along the depth for an eroded specimen in LiBr solution

The microhardness profile along the depth is depicted in Figure 3-21. After 60 min of test, work-hardening layer with a thickness of 70 μm from the surface to substrate can be observed, and the hardness is increased from330 MPa to 370 MPa. As the test is progressed to 180 min, the work-

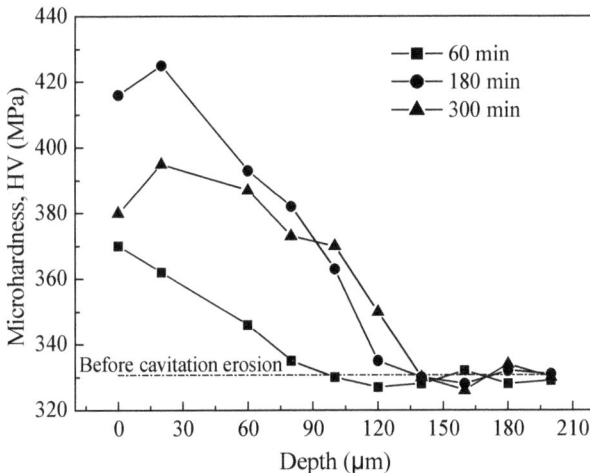

Figure 3-21 Microhardness versus depth in the cross section of Ti-6Al-4V after cavitation erosion in LiBr solution for different time

hardening layer is thickened and it is 120 μm. The hardness at a depth of 20 μm from the surface reaches 416 MPa, which is increased by 26% in contrast with that of the specimen prior to cavitation erosion. After 300 min, the hardness declines. But the work-hardening layer is thickened to 140 μm. For the sub-surface in the range of 60-100 μm, the hardness has the value above 370 MPa, which is higher than that near the surface (40 μm). This indicates a work softening phenomenon.

The hardness test results can give us some understanding of the cavitation erosion process. In the initial stage of cavitation erosion, work-hardening results from the impinging induced by bubble collapse. As the cavitation erosion continues to the end of the incubation period, the hardness near the surface is increased and reaches the maximum value. Afterwards, the hardness goes down. A great number of surface materials are removed, causing work softening near the surface. In the meantime, the work-hardening layer moves deep into the substrate. The rate of material removal near the surface is lower than the rate of moving high hardness layer deep into the substrate, which explains the thickening of work-hardening layer and the appearance of work softening region with a certain thickness near the surface.

2) Residual stress

An obvious residual stress increase on the surface of a material may be oberved during the cavitation erosion process. This indicates the strong impacts caused by cavitation bubble collapses. Figure 3-22 shows the plot of residual stress versus test duration. Before the cavitation erosion test, the residual stress of the polished specimen is around 130 MPa. After 20 min, the residual stress rises sharply to 737 MPa. Then it levels off.

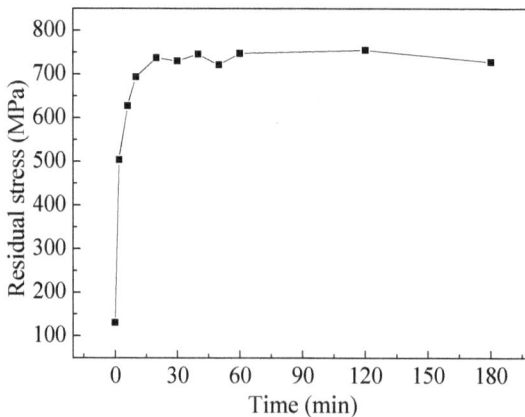

Figure 3-22 Residual stress on the surface of Ti-6Al-4V after different cavitation erosion duration in LiBr solution

Since no extra stress is produced by chemical etching, the eroded specimen was chemically etched to different depths, so that the profile of the residual stress along the depth can be obtained (Figure 3-23).

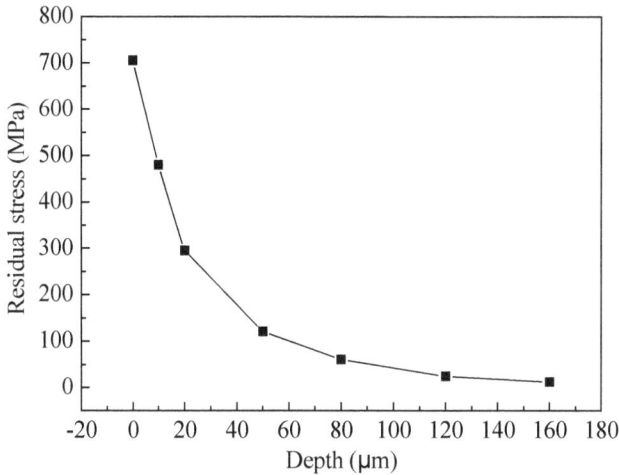

Figure 3-23 Residual stress versus depth after cavitation erosion for Ti-6Al-4V in LiBr solution for 180 min

From the data in Figure 3-22, the residual stress on the surface is 706.2 MPa after 180 min of the test, which suggests that serious plastic deformation causes the dislocation pile-up, leading to an increase in dislocation density and the lattice distortion, ultimately generating a great residual stress. With an increase of depth into the substrate, the residual stress drops rapidly. At a depth of 120 μm, the residual stress is only 25.3 MPa, which can be nearly ignored. The results show clearly that the thickness of the stress layer is within 120 μm in the end of the incubation period. The mechanical properties in the deeper layer of substrate are not affected by cavitation erosion. This agrees with the result of the cross-sectional microhardness measurement.

2D and 3D AFM images of Ti-6Al-4V after different periods of cavitation erosion test are shown in Figure 3-24.

The plastic deformation occurs on the surface due to the cavitation (Figure 3-24b). At 20 min, the residual stress on the surface of Ti-6Al-4V reaches a peak and the plastic deformation is aggravated (Figure 3-24c). In this period, no material on the surface falls off. At 60 min, Figure 3-24d shows noticeable plastic deformation on the surface. The surface present a different appearance due to spalling of surface materials.

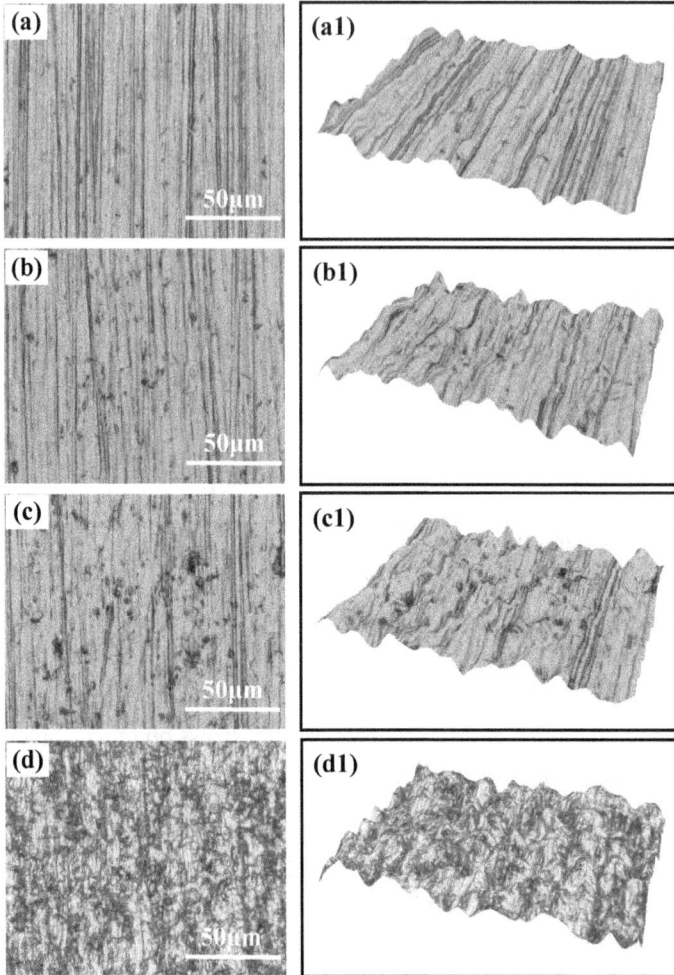

Figure 3-24 2D and 3D images of Ti-6Al-4V after different cavitation erosion duration in LiBr solution: (a) 0 min, (b) 6 min, (c) 20 min, (d) 60 min

3.4.3.5 Electrochemical Corrosion Behavior

In the above investigations, electrochemical corrosion has been shown to play a role in the overall degradation of titanium alloy in LiBr solution during cavition erosion process. In this section, electrochemical behaviors under static and dynamic conditions are discussed. The techniques, open circuit potential measurement (OCP), potentiodynamic polarization and electrochemical impedance spectrum (EIS), provide useful information with regards to the breakdown and repassivation of the passive film, corrosion rate and sensitivity of specific materials to corrosion in designated environments.

1) Open circuit potential

The OCP versus time curve of Ti-6Al-4V in LiBr solution under alternating conditions of quiescence and cavitation is given in Figure 3-25.

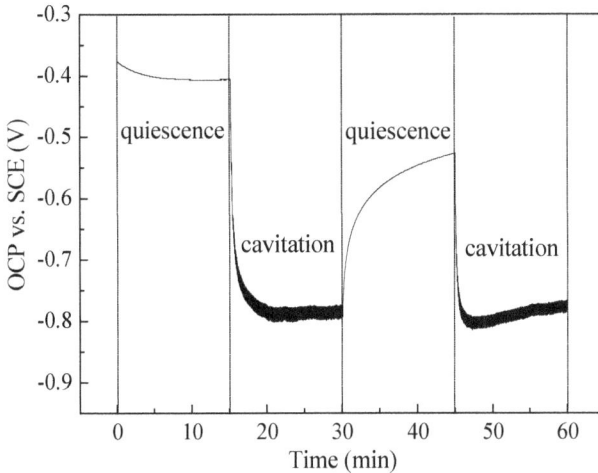

Figure 3-25 Open circuit potential versus time of Ti-6Al-4V in LiBr solution under alternating condition of quiescence and cavitation

Under static condition, the OCP of Ti-6Al-4V is about -400 mV vs. SCE. Once the electrode is subjected to cavitation, the OCP rapidly shifts towards more negative potential by 300 mV. This can be explained by the destruction of the passive film covering on the surface and exposure of the fresh metal surface due to continuous shock waves and micro jets caused by the collapsing of bubbles. When cavitation stops, the OCP gradually moves in the positive direction, indicating the occurrence of repassivation of the passive film without the impingement of mechanical force. When cavitation is present again, the OCP shows the rapid negative shift again.

The morphological inspection of the electrode surface after the 60-minute test was done by optical microscope (Figure 3-26). A small quantity of pits can be found on the surface. This reveals that the metal surface is actively corroded in small areas. The repassivation under static condition can impede the damage to some extent.

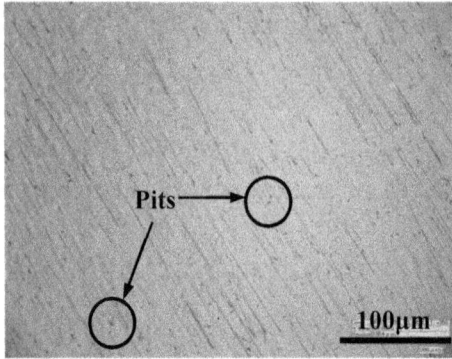

Figure 3-26 Image of the electrode surface after 60-minute OCP test under cavitation condition

2) Potentiodynamic polarization and EIS curves

Potentiodynamic method is used to produce the overall shape of the polarization curve, revealing corrosion form and passive ability of a material in the particular medium. Cyclic potentiodynamic polarization tests are often used to evaluate pitting sensitivity and repassivation behavior.

The potentiodynamic polarization curves of Ti-6Al-4V in distilled water and LiBr solution under static and cavitation conditions are shown in Figure 3-27. Table 3-6 lists the value of electrochemical parameters

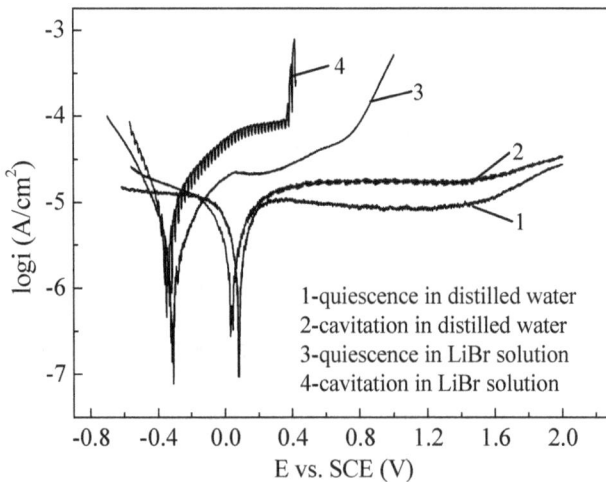

Figure 3-27 Potentiodynamic polarization curves for Ti-6Al-4V in distilled water and LiBr solution under static and cavitation conditions

under both conditions, such as corrosion potential E_{corr}, corrosion current density i_{corr}, maintaining passive current density i_p and passive region $(E_{tp}$-$E_p)$.

Table 3-6 Electrochemical parameters for Ti-6Al-4V in distilled water and LiBr solution under static and cavitation conditions

Solution	Condition	E_{corr} (mV)	i_{corr} ($\mu A \cdot cm^{-2}$)	i_p ($\mu A \cdot cm^{-2}$)	Passive Region $(E_{tp}$-$E_p)$(mV)
LiBr	Static	-320.5	2.922	22.338	655.5
	Cavitation	-352.4	5.322	78.363	271.4
Distilled water	Static	79.1	2.735	10.599	1204.1
	Cavitation	36.6	3.302	17.449	1025.6

Under static condition, in distilled water the corrosion potential is 79.1 mV. The passive zone has a wide potential range, and the difference between maintaining passive potential and transpassive potential is 1204.1 mV. In this passive zone, the current has a relatively stable value independent of potential. In LiBr solution, the corrosion potential shifts in the active direction, corrosion current density increases and the range of the passive zone decreases to 655.5 mV. At potentials about 1400 mV, the current density starts to go up. Based on the research of other authors (García-García 2006, García-García et al 2008), this is associated with the localized oxidation of bromides ions to bromine at electroactive sites and to the subsequent formation of HBrO:

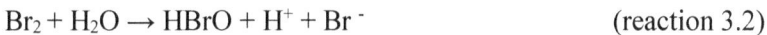

$$2Br^- \rightleftharpoons Br_2 + 2e^- \qquad \text{(reaction 3.1)}$$

$$Br_2 + H_2O \rightarrow HBrO + H^+ + Br^- \qquad \text{(reaction 3.2)}$$

Under cavitation condition, the shape of the polarization curve in distilled water is similar to that under static condition. But as compared with static condition, the corrosion potential is more negative by 42.5 mV, and the passive zone becomes narrower. The maintaining passive current density oscillates, suggesting that the passive film on the Ti-6Al-4V electrode surface is in a metastable state of depassivation/repassivation. The cavitation can damage the passive film or make it thinner. Whereas in LiBr solution, the corrosion potential has a significantly negative shift, the corrosion current density and maintaining passive current density are remarkably increased. The passive zone has a decreased range of 271.4 mV. Moreover, in contrast with the electrochemical test results in distilled water, the maintaining passive current density has the bigger oscillating amplitude. The transpassive potential decreases significantly. The current density shows a sharp increase at potentials higher than the transpassive potential. The active dissolution of titanium, rather than Br⁻ oxidation, may explain the rise in the current density.

In addition, it can be seen from the cathodic polarization curves that cavitation can bring some impact on the cathodic reaction. Under static condition, the cathodic reaction is controlled by the diffusion process. There is a limiting diffusion current density, which is independent of the applied potential. The presence of cavitation can stir the solution, accelerating the diffusion of corrosion species and corrosion products. The rate determining step is converted from the diffusion process to the electrochemical reaction.

Cavitation has more influence on the anodic branch than the cathodic branch. The E_{corr} value is more negative under cavitation condition than under static condition, and the i_{corr} and i_p values are increased by the effect of cavitation.

Potentiodynamic polarization and EIS in different cavitation periods can be used to discuss the effect of cavitation duration on the electrochemical behavior of Ti-6Al-4V. The curves of polarization are depicted in Figure 3-28. The corresponding electrochemical parameters are given in Table 3-7.

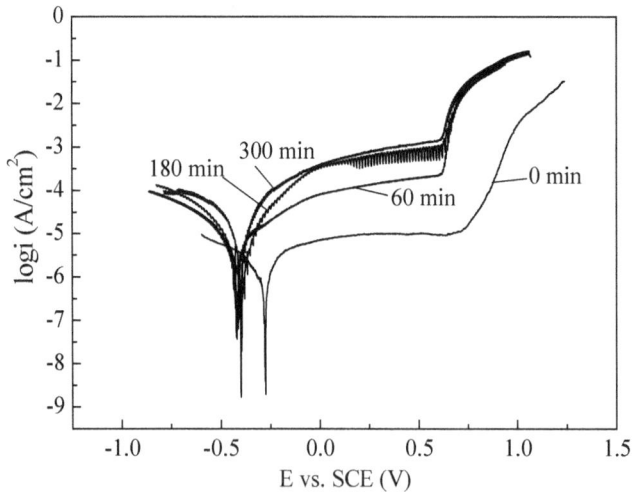

Figure 3-28 Potentiodynamic polarization curves for Ti-6Al-4V in LiBr solution during different cavitation period

Table 3-7 Electrochemical parameters for Ti-6Al-4V in LiBr solution during different cavitation period

Period (min)	i_{corr} (A/cm^2)	i_p (A/cm^2)	v (g/(m^2·h))	R_p (Ω·cm^2)
0	5.034×10^{-7}	9.195×10^{-6}	2.25×10^{-3}	2.98×10^5
60	3.382×10^{-6}	1.152×10^{-4}	2.16×10^{-2}	5.12×10^4
180	4.687×10^{-6}	8.078×10^{-4}	2.67×10^{-2}	3.56×10^4
300	8.753×10^{-6}	1.228×10^{-3}	3.62×10^{-2}	2.93×10^4

After 60 min, the polarization curves recorded show some changes compared to the curve of Ti-6Al-4V without undergoing cavitation. The i_{corr} and i_p values sharply increase. The abatement of the passive region range and oscillation of the passive current density can be found. At 180 min, this corresponds to the end of the incubation period. There is a distinct oscillation of the passive current density in the anodic branch. Visual examination of the electrode surface indicates the formation of some pits. In the polarization curve for the cavitation test of 300 min (acceleration period), there are larger i_{corr} and i_p values.

Figure 3-29 shows the morphological inspection results of the electrode surface after the test. Corrosion pits grow on the electrode surface after anodic polarization. The number and size of pits increase as the erosion time increase. The cavitation action leads to a significant reduction of the passive range. When the applied current is polarized to the transpassivation zone, the passive film is rapidly destroyed by the synergetic effect of cavitation and electrochemical corrosion. The repassivation rate of the passive film is less than the corrosion dissolution rate, which enables corrosion to occur. The longer the cavitation erosion time is, the greater the damage degree is. The mechanical properties are degraded in the incubation stage and the activity of the localized surface is strengthened, causing a sharp rise in the volume and number of pits.

Electrochemical impedance spectrum (EIS) is a useful technique for characterizing the performance of passive film on the surface of titanium alloy in the process of caitation erosion. Figure 3-30 is the Nyquist diagrams of Ti-6Al-4V after different periods of cavitation erosion.

EIS for Ti-6Al-4V unaffected by cavitation shows a capacitive arc with a large diameter. It suggests that the Ti-6Al-4V surface has good resistance because of the existence of tenacious and adherent oxide film. In the presence of cavitation, the mechanical properties of the surface are changed, and the impedance radius becomes smaller. At this time, the

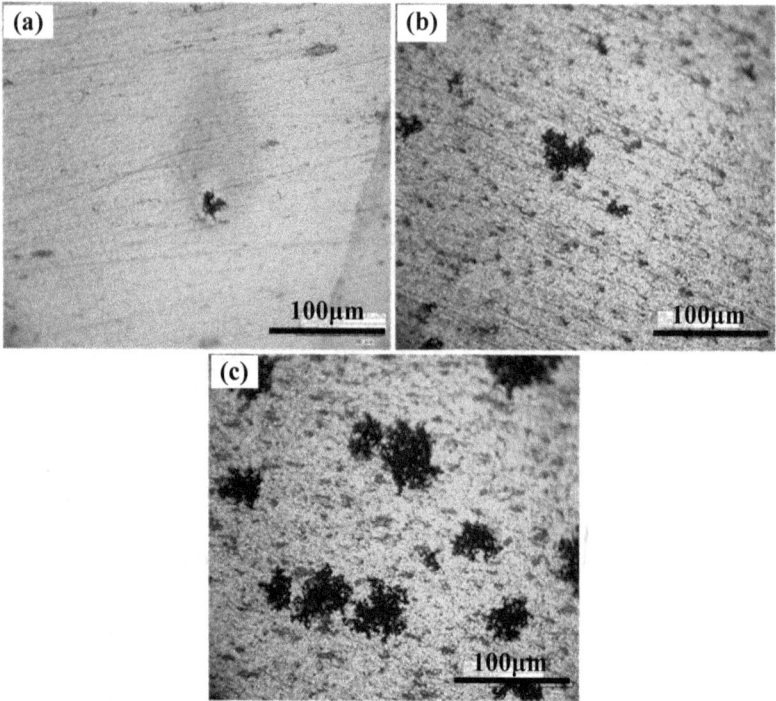

Figure 3-29 Images of Ti-6Al-4V electrode surface after polarization test under cavitation condition: (a) 60 min; (b) 180 min; (c) 300 min

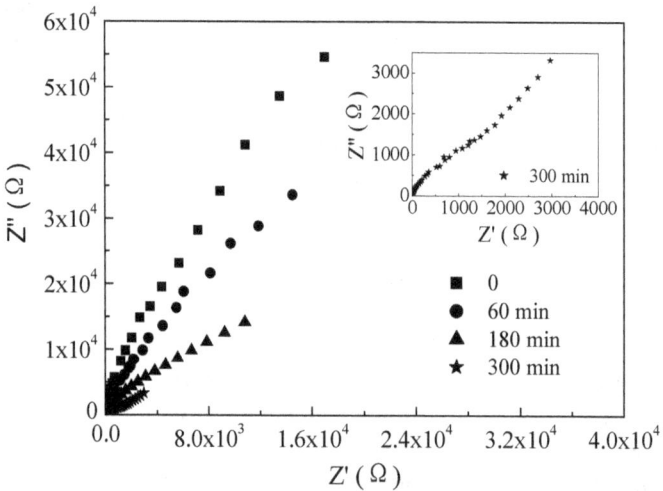

Figure 3-30 Impedance characteristics of Ti-6Al-4V under cavitation erosion during different cavitation period

electrode is not only affected by the electrode potential, but also affected by the destruction of passive film on the surface. The Nyquist diagram comprises of a high-frequency capacitive arc and a low-frequency linear Warburg impedance after 180 min. Br⁻ absorbs on the surface with the damaged passive film that prevents the growth of the passive film and accelerates the thinning of the passive film. Localized corrosion occurs on the electrode surface. So, the reaction resistance decreases continuously and the corrosion rate increases with cavitation time.

Through the analysis and fitting of the impedance spectra, the equivalent circuit is obtained and shown in Figure 3-31. In the figure, R_s represents the solution resistance between the working and the reference electrodes; R_{ct} represents the reaction resistance; C is the equivalent capacitance, representing interfacial capacitance and Z_w is Warburg resistance caused by the diffusion process. The corresponding value of each component is given in Table 3-8. Under the condition of cavitation, the reaction resistance (R_{ct}) decreases continuously due to the stirring effect induced by rapid formation and collapse of bubbles and further reduction of resistance to oxygen diffusing towards the electrode surface. R_{ct}, as charge transfer resistance, is closely associated with the anodic dissolution process. Its decrease indicates the rise in dissolution rate during the cavitation process. This coincides with the results of potentiodynamic polarization. As the cavitation progresses, the deformation and corrosion in large areas occur, resulting in the increase of the corrosion specific surface area. This is manifested by the increase in the interfacial capacitance.

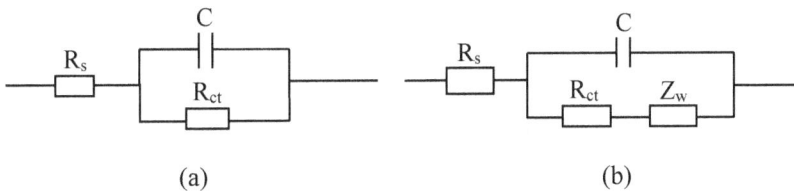

(a)   (b)

Figure 3-31 Equivalent circuit of impedance of Ti-6Al-4V under cavitation erosion during different cavitation period: (a) 0 and 60 min; (b) 180 and 300 min

Table 3-8 EIS parameters in the equivalent circuit

Time (min)	0	60	180	300
R_s /$\Omega\cdot cm^2$	0.983	0.635	0.7495	0.401
C/F·cm²	1.608×10^{-5}	2.688×10^{-5}	3.152×10^{-5}	4.762×10^{-5}
R_{ct} /$\Omega\cdot cm^2$	2.988×10^{5}	2.757×10^{4}	1758	1279
Z_w/$\Omega\cdot cm^2$	—	—	2.233×10^{-5}	2.886×10^{-4}

3.4.3.6 Discussion of Cavitation Erosion Mechanism

1) Synergetic effect of cavitation and corrosion

① Qualitative analysis

The cavitation damage is caused by the combination of mechanical and corrosion effects in the corrosive medium. The combination of the two effects has a greater impact than each individual damage to the material.

Promoting effect of cavitation erosion to corrosion

When mechanical effects impact the titanium surface, a high stress concentration is generated in a small region. This causes a local plastic deformation of the surface. There is a large difference in internal energy between the material in the deformed region and the un-deformed region, creating electrochemical microcells on the surface. This dramatically increases the reactivity of the surface and accelerates the process of corrosion. Afterwards the high-velocity impingement during the collapse of bubbles further induces the thinning or breakdown of the protective passive film and the production of micro-cracks, thereby exposing the fresh surface. More electrochemical microcells form as a result of the difference in electrochemical properties on the surface. The high and repeated impact enables the electrochemical microcells to keep developing. Simultaneously, cavitation stirs the solution, increasing the number of reactants supplied to the surface and enhancing the cathodic reaction rate. During this process, repassivation occurs, but the re-growth rate of the passive film is less than the corrosion dissolution rate. Apparently, this constitutes a vicious cycle, which results in the formation of numerous cavitation pits, the damage and loss of surface materials.

Promoting effect of corrosion to cavitation erosion

The roughening of the surface caused by the corrosion and the emergence of localized corrosion result in more stress concentration points. This produces the perfect condition for the development of cavitation erosion, which furthermore decreases the mechanical properties on the surface and promotes the process of cavitation.

② Quantitative analysis

Electrochemical analysis was used in conjunction with mass-loss analysis to determine the total material loss and to isolate the contribution due to pure corrosion.

For the cavitation erosion system in this research, the cumulative mass loss W_T is composed of three parts: pure cavitation W_e, pure corrosion W_c and combined action of corrosion and cavitation W_s.

$$W_T = W_e + W_c + W_s \qquad (3.3)$$

The synergetic effect of mechanical action and corrosion also contains two parts: the contribution of corrosion to cavitation ΔC and the contribution of cavitation to corrosion ΔE.

$$W_s = \Delta C + \Delta E \qquad (3.4)$$

Hence Eq.(3.5) can be rewritten as

$$W_T = W_e + W_c + \Delta C + \Delta E \qquad (3.5)$$

W_e can be obtained from the mass loss in distilled water. W_c and ΔC can be calculated by the corrosion current density in LiBr solution under static condition and dynamic condition, respectively. ΔE can be calculated from Eq. (3.5). The results during different periods are summarized in Table 3-9. The percentage of mass loss induced by pure corrosion during various periods is less than 0.15%, whereas the contribution of the combined action accounts for 10.96% - 21.49%. Pure mechanical action results in damage of over 78%. This verifies the leading role of the mechanical effect. However, in the synergetic contribution, corrosion shows the obvious promoting effect to cavitation erosion.

Table 3-9 The percentage of mass loss induced by pure corrosion, pure cavitation and combined action

Time (min)	W_T (mg)	W_e/W_T (%)	W_c/W_T (%)	$\Delta C/W_T$ (%)	$\Delta E/W_T$ (%)	W_s/W_T (%)
60	0.53	79.24	0.13	2.07	18.56	20.63
180	1.84	88.04	0.12	1.85	9.99	11.84
300	4.17	88.96	0.08	1.36	9.60	10.96
480	6.86	78.43	0.08	1.31	20.18	21.49

2) Cavitation erosion behavior

At the initial stage of cavitation erosion of Ti-6Al-4V, plastic deformation preferentially occurs in the low strength phase (α phase) which absorbs the impact energy generated by bubble collapse, therefore plastic deformation degree in the high strength phase (β phase) is reduced. This results in heterogeneous deformation on the surface. In the meantime, the internal energy of α phase on the surface of Ti-6Al-4V is increased rapidly, and its potential is lower than that of the surrounding area, which leads to the inhomogeneity of electrochemical property on the surface. α phase is more likely to suffer from cavitation damage. On the α phase, the breakdown of the passive film makes the fresh surface exposed, forming small anodes and large cathodes. LiBr medium reacts with the substrate, causing the active dissolution in the regions of α phase and then the formation of pits. However, the work-hardening is also yielded, which improves the resistance to plastic deformation. This suggests that the increasing rate of surface roughness Rq is slightly decreased at the end of the initial stage. As the cavitation erosion continues, the corrosion pits increase the damage probability. The local stress is more concentrated and the mechanical factor is strengthened. Moreover, the agitation of cavitation promotes the diffusion of corrosion products into the solution and increases the migration of corrosive species to the surface. The synergetic effect of mechanics and electrochemical corrosion results in the development of the existing pits and emergence of new pits. Eventually β phases distributed at the boundary of pits fall off and the degree of surface concave and convex becomes small. Figure 3-32 is a graphical representation of cavitation erosion process at the initial stage of Ti-6Al-4V.

Although the cavitation erosion depth in different areas on the surface of Ti-6Al-4V varies with randomness of cavitation phenomenon and microstructure, the damage of cavitation erosion propagates along the vertical direction and surface profiles in different areas are similar (Figure 3-33). The gray line is the surface profile curve before the cavitation

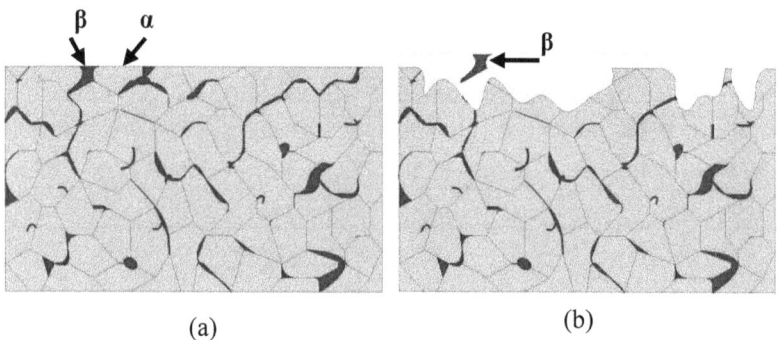

(a)   (b)

Figure 3-32 Schematic representation (a) before and (b) after cavitation erosion of Ti-6Al-4V for 180 min

erosion. In the steady-state stage, the cavitation erosion damage proceeds and reaches a dynamic balance. The increasing rate of average depth of cavitation erosion is significantly improved, while the surface morphology basically remains invariant.

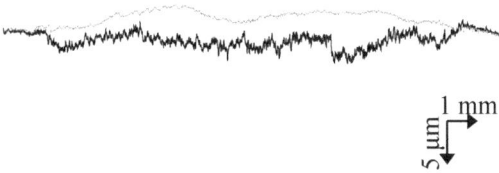

Figure 3-33 Surface profile of cavitation erosion of Ti-6Al-4V after 900 min

3.4.4 Cavitation Erosion Behavior of Commercially pure Ti

A CP-Ti was used for comparison of the cavitation erosion resistance with Ti-6Al-4V. The differences of the microstructure of CP-Ti from that of Ti-6Al-4V are (Figure 3-34):

1) The constitution of CP-Ti is α phase;

2) It has larger grain size.

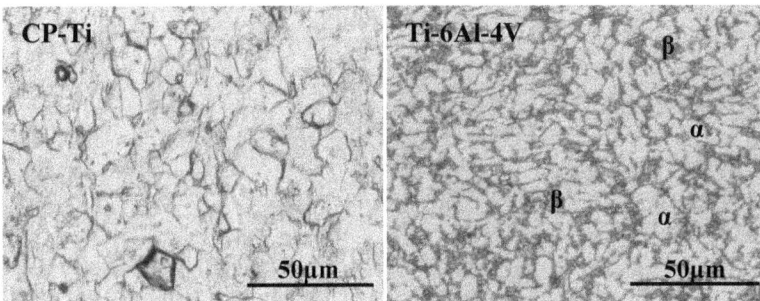

Figure 3-34 Microstructure of CP-Ti and Ti-6Al-4V

3.4.4.1 Surface Roughness and Mean Depth of Erosion

Three stages are found in the evolution of surface roughness and mean depth of erosion, incubation, accumulation and stationary, see Figure 3-35. This is similar to that of Ti-6Al-4V. Table 3-10 gives the duration of the three stages for Ti-6Al-4V and CP-Ti. The result indicates that CP-Ti has a shorter incubation time.

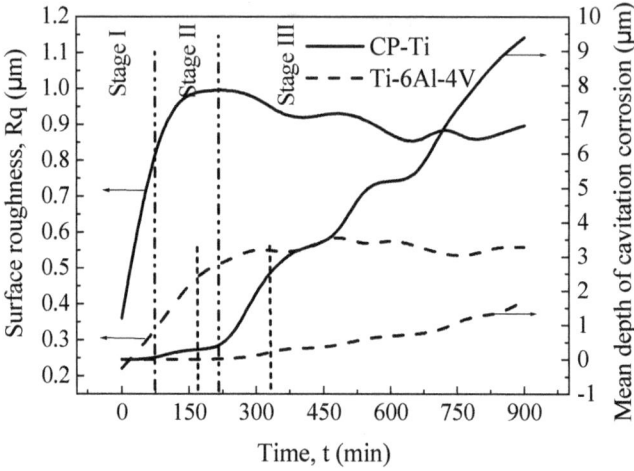

Figure 3-35 Variation of surface roughness and mean depth of erosion for CP-Ti

Table 3-10 Three stages of cavitation erosion process for Ti-6Al-4V and CP-Ti

Stage	CP-Ti	Ti-6Al-4V
Incubation	0-160 min	0-70 min
Accumulation	160-320 min	70-210 min
Stationary	>320 min	>210 min

3.4.4.2 Surface Morphologies

Figure 3-36 shows the morphologies of the eroded surface of CP-Ti after different periods. Figure 3-37 shows the images of cavitation erosion of CP-Ti etched before the test.

The surface produces uneven plastic deformation, and as the cavitation test continues, the degree of plastic deformation of CP-Ti and the dislocation density of grain boundary increase. This causes the concentration of stress, slight spalling of materials on the grain boundary, and the production of local cracks. The surface roughness value (Rq) increases linearly with time. With the extension of cavitation test, the cracks propagate along the grain boundary or into the grains, the spalling

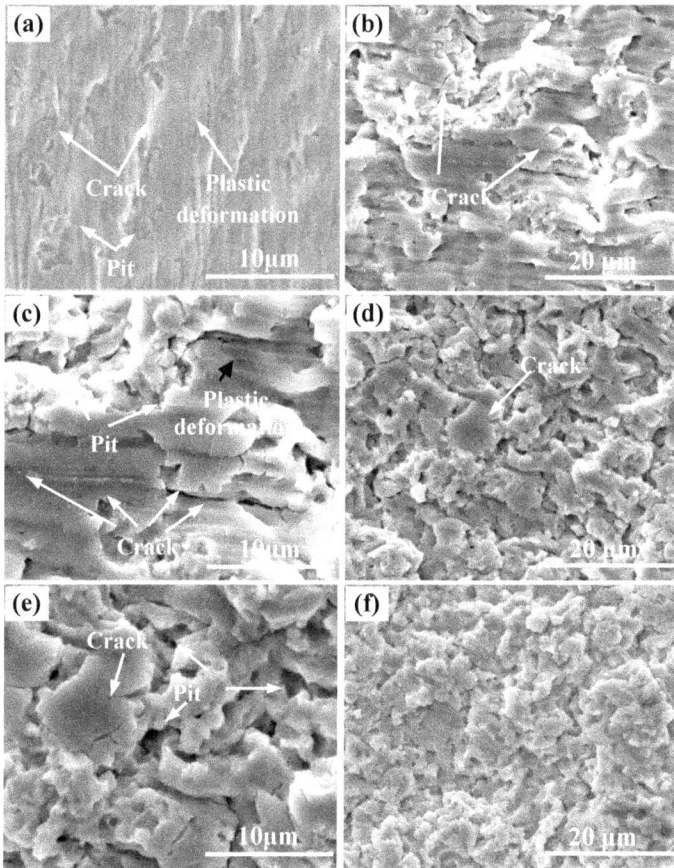

Figure 3-36 Morphologies of CP-Ti after cavitation erosion test in LiBr solution:
(a) 30 min; (b, c) 60 min; (d, e) 120 min; (f) 180 min

of materials increases, at the same time, the passive film on the surface is easily attacked and fresh substrate is exposed. This promotes the corrosive effect of LiBr. Then the synergistic effect of cavitation and corrosion accelerates the growth of corrosion pits. Eventually the materials at the boundary of pits fall off, and the degree of surface concave and convex become smaller. The growth rate of Rq decreases and stabilizes at the end. This further proves that the mechanical action is the dominant factor of cavitation erosion damage of titanium alloys in LiBr solution, and electrochemical corrosion has the accelerating impact.

The resistance of cavitation erosion of Ti-6Al-4V is superior to CP-Ti. This may result from the microstructure and grain size of materials. For Ti-6Al-4V, the relatively small grain size and the existence of high strength β phase reduce the concentration of stress on the surface and prevent the cracks from further propagating.

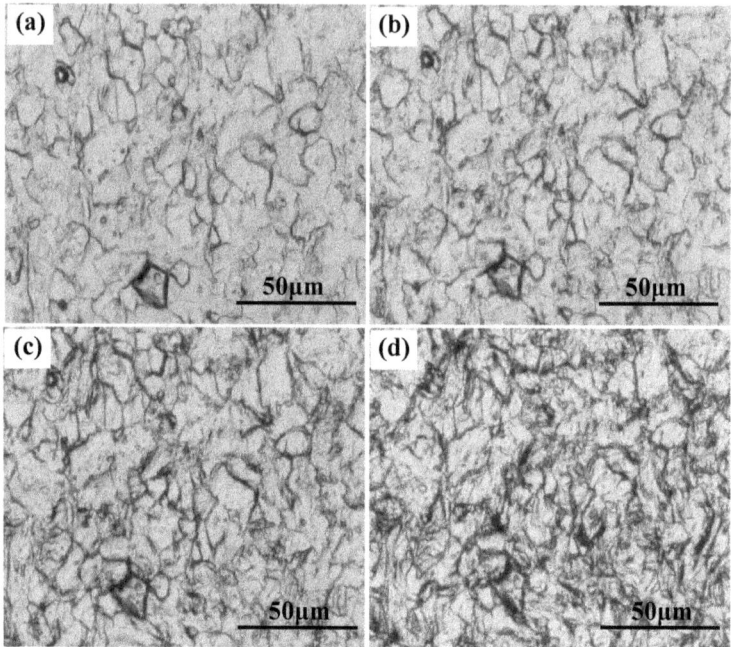

Figure 3-37 Influence of microstructure of CP-Ti on cavitation erosion observed by using three-dimensional video microscopy: (a) 15min; (b) 30min; (c) 45min (d) 60min

Chapter 4 - Chemical Etching

4.1 History

Corrosion is a destructive phenomenon which affects almost all metals. If the corrosion process is effectively controlled, it can be applied as a method for cutting materials. This process is called "chemical etching" (this term will be used in this book). It is also named "chemical milling" and "chemical machining".

Chemical etching is a nontraditional processing method, which is employed to carry out the precision contouring of metal into any size, shape or form and to remove unwanted parts at a predetermined location, extent and depth from the workpiece material by a controlled chemical reaction, without the use of physical force. The two key materials used in the chemical etching process are etchant and maskant (Wikipedia-chemical milling 2016).

It is difficult to use ordinary water in a short period of time to chemically process a metal surface. It is necessary to have a corrosive substance - acid. Organic acids, such as lactic acid extracted from yogurt, citric acid and acetic acid extracted from lemon, were used first. But these organic acids were not enough to complete chemical dissolution of a metal. It was not until inorganic acid was discovered that metal chemical etching became possible. However chemical etching technology lagged behind the discovery of inorganic acids, which may be related to the development of effective protective coating (maskant).

Metal chemical etching originated in the 15th century in Medieval Europe and was used aesthetically to decorate metal armor, cups and other metal objects. In this period, plate armor was very common. Craftsmen often applied the paint to various parts in plate armor. There were two purposes, one was to identify the armor as a marker and the other was to prevent atmospheric corrosion. Perhaps, accidental damage of paint led to corrosion of local areas in the armor. The rust looked like a man-made pattern. It was likely this phenomenon caused the craftsmen to then purposefully make numerous types of markers in the armor. There was another possibility: in those days, the metallic parts for human body protection, such as weapon, helmet, chestplate, etc., made by a forging method, had high hardness, resulting in a difficult application of commonly adopted hand engraving. This made artists explore new ways of pattern formation. Through a variety of tests, an acid corrosion processing method was found and became an alternative to engraving (Harris 1976).

The earliest written record about chemical etching described a maskant of linseed-oil paint applied to protect the areas where corrosion was not needed and a chemical etchant mix of salt, charcoal and vinegar used to bite into the unprotected areas. During the process, the first step was to paint the protective coating according to the designed configuration. After drying, the etchant was used to corrode unprotected surfaces. At a later time, natural resins, such as paraffins, were applied as a protective layer. In this method, the whole surface of the metal was coated with a paraffin wax, and the part of the protective layer which was required to be corroded was removed in advance. As we all know, this is the origin of selective chemical etching.

From the 17th to 19th century, the discovery of a variety of acids and alkalis favored the further development of chemical etching technology (Yang 2008). Johann Rudolf Glauber (The Chemical Engineer 2012), a German alchemist, invented a new method of producing hydrochloric acid, in which seawater and sulfuric acid were mixed to distill and condensed products (enriched hydrochloric acid) were collected into a container filled with water. The etching rate of iron in hydrochloric acid was fast, which enabled the chemical corrosion efficiency to be significantly improved. Basil Valentine (Harris 1976), also a German alchemist, studied various kinds of corrosive agents. He found that caustic alkali (generated by the reaction between caustic lime and Na_2CO_3 or K_2CO_3 with medium concentration) was a potent liquid. No doubt, this led to the widespread application of caustic alkali in chemical etching of aluminum today. In 1771, hydrofluoric acid was discovered, which is of great significance to the chemical etching of titanium, tungsten and other high-temperature corrosion resistant materials today. At this stage, the method of producing sulfuric acid was greatly improved. Sulfur and potassium nitrate were mixed to prepare sulfuric acid by combustion. This completely replaced the method of sulfate calcination. In the same period, Nicolas Leblanc from France invented a method of turning salt into carbonate, which made hydrochloric acid become one of the most inexpensive inorganic acids at that time. These improvements provided more choices for metal chemical etching.

In the early 20th century, the development of stronger resists allowed more highly corrosive etchants to be used on stronger metals. Around the 1940s, chemical etching was widely used to etch thin samples of very hard metal. The main industrial application of chemical etching was developed after the Second World War. In 1953, North American Aviation Inc. used the chemical dissolution to etch aluminum components for rockets. (US

Patent No: 2739047, 1956). In the mid-60s of the 20th century, the application of chemical etching greatly was increased in both quantity and quality. In the aviation and aerospace industries, a reliable process was desired for removing excess metal on pre-formed parts, such as a missile shell, and a number of aircraft components. Obviously, chemical etching was the best way to solve this problem. In the manufacturing of flight vehicles, chemical etching could not only reduce the weight of the parts, but most importantly also produce various configurations, such as grooves, slots, etc., in bent thin wall metal materials. In some cases, this would have been done difficultly through traditional physical machining.

Currently, chemical etching is a processing method with very practical value. It can enable many parts to be produced simply, easily and inexpensively compared with other conventional machining methods. It is mainly applied in the following facets:

- Reducing the weight of parts,
- Processing thin-wall parts,
- Producing geometrically complex and precision parts.

More information about the history of chemical etching can be found in the book "Chemical milling-the technology of cutting materials by etching" by Harris (1976) and the book "Metal etching technology" by Yang (2008).

4.2 Characteristics

Chemical etching is fundamentally different from conventional machining methods (Shen 1984, Cakir 2005). During the process of chemical etching, no mechanical force acts on the cutting surface. In addition, under the effective protection of maskant, metal materials in the specified areas can be removed delicately and precisely, and simultaneously the material thickness prescribed by the drawing in the areas where necessary can be maintained. The process is simple to implement and has low production cost, low surface roughness and high processing precision. Inherent advantages of chemical etching over conventional metal working methods include:

1) It is easy to process sheets that are large, thin, and easily deformable.

2) It is not affected by the state of metal material, regardless of if it is formed, heat-treated, hardened.

3) Some welding, riveting, bonding components can be changed to the integral structure, which can be machined directly using chemical etching with the purpose of reducing structural weight, reducing equipment, saving working hours, and shortening the production cycle.

4) No stress is introduced to the workpiece, which minimizes part distortion and makes processing of delicate parts possible.

5) The good surface quality eliminates the need for subsequent finishing operations.

6) A number of parts can be processed at one time and both sides of the parts can be processed simultaneously. These lead to high productivity.

7) It is easy to control the rate of dipping in or out of the container for parts.

8) It can save material in comparison with a mechanical machining process.

9) Design changes can be implemented quickly. When the part design is changed, it is only necessary to modify the template, which can be applied to the trial manufacture of new parts.

10) The equipment is relatively simple and the capital cost is relatively low.

However, there are some limitations for chemical etching:

1) It is a successive cutting process based on the original surface state of part, so that the shape and surface state of the part after chemical etching are directly affected by the original ones.

2) It is difficult to form sharp corners (Figure 4-1a). The shape of the edge after chemical etching is a circular arc with a radius approximate to the etching depth (Figure 4-1b).

3) It is difficult to work in narrow and deep cavities. The bubbles produced from the chemical reaction are stuck under the edge of the maskant, and prevent the etchant from contacting with the metal substrate (Figure 4-2). This results in irregular corrosion and forms irregular edges. Some soft maskants can enable the bubbles to come out easily, but after a certain depth, even the use of mechanical stirring is not enough to make the bubbles come out from the edge of the maskant. The most effective way of solving this problem is to apply a time-consuming manual method of trimming the contour edge of the

maskant. Another problem is the effect of surface tension of the etchant, which makes it difficult to work on narrow or small-radius surfaces. For deeper cavities, the width should be greater than 4 mm, while for shallower cavities, width or radius are not less than 1.5 times the depth.

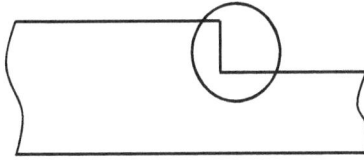

(a) Edge with sharp corner that can not be processed by chemical etching

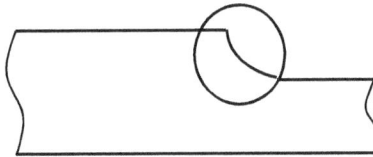

(b) Edge that is processed by chemical etching

Figure 4-1 Limit of geometry of edge after chemical etching

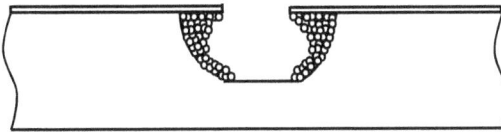

Figure 4-2 Bubbles stuck under the edge of maskant

4) Hand masking, scribing, and stripping can be time-consuming.

5) Hydrogen pickup and intergranular attack are problems with some materials.

6) Etchants may be very dangerous for workers and disposal of etchants is expensive and troublesome.

4.3 Classification and Applications

Based on the requirement for the parts to be processed, chemical etching can be classified as (Figure 4-3) (Shen 1984):

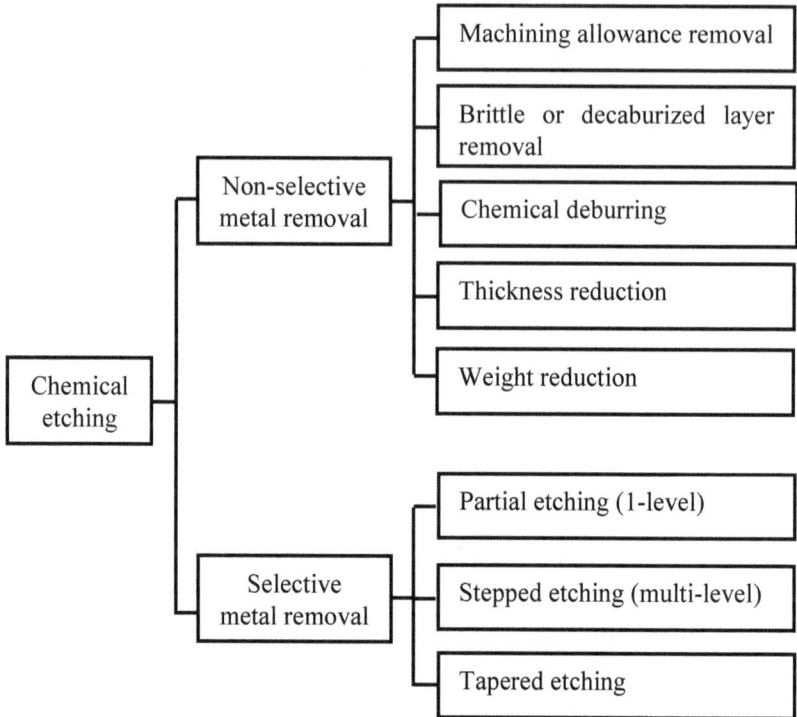

Figure 4-3 Classification of chemical etching

1) Non-selective metal removal

In this process, the workpiece uncoated with maskant is immersed into the solution and all surfaces are uniformly milled (Figure 4-4a). This is mainly used for removing machining allowance, brittle layers (alpha case) on the surface of the semi-finished products after precision forging, decarburized layers on the surface of low alloy steel forgings, and sharp burrs in conventional machined parts. It is also a means of thickness reduction (such as processing the wall of aluminum castings) and weight reduction.

2) Selective metal removal

In this process, maskant is applied to non-etching sections of the workpiece and then chemical etching is used to process unprotected sections. This is used for manufacturing complex shape workpieces, such as large thin walls, honeycomb structure panels, welding assemblies, integral strengthening walls, aircraft skins, airframes, electronic instrument circuit boards, and so on.

In most cases, workpiece thickness can be reduced locally and contour surface can be obtained by 1-level selective etching (Figure 4-4b). Some workpieces may however need multi-level selective etching to obtain the required thickness (Figure 4-4c).

Chemical etching can be applied in manufacturing different kinds of materials, including metals and non-metals:

– Metal

Aluminum, copper, zinc, steel, lead, nickel, titanium, molybdenum and zirconium

– Non-metal

Glass, ceramics and plastics

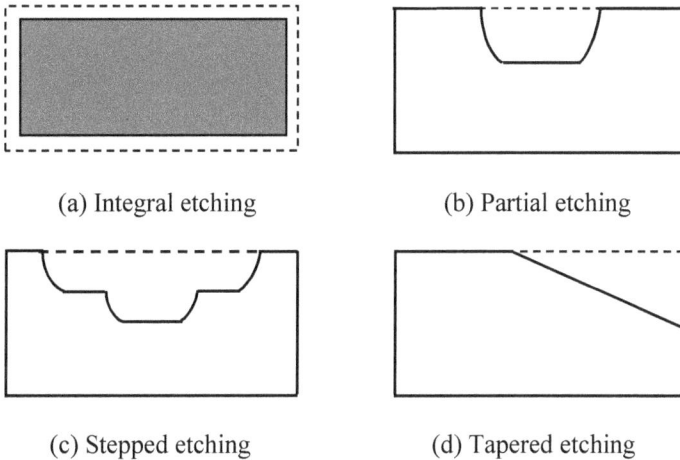

(a) Integral etching   (b) Partial etching

(c) Stepped etching   (d) Tapered etching

Figure 4-4 Chemical etching configuration

Chemical etching has broad applications in the aerospace industry. It is also used in the civil industry.

1) Application in the aerospace industry

In airplanes, missiles, satellites, spaceships, astronaut cabins, gyroscopes, heat exchangers, rockets, etc., chemical etching, as a non-conventional processing method, has become a popular technique of removing shallow layers from large components and producing cavities on plates, sheets, forgings and extrusions.

With the rapid development of the aerospace industry, the importance of chemical etching technology has attracted more and more attention for various purposes:

- Reducing structure weight;
- Shaping taper workpieces;
- Processing thin-wall workpieces;
- Producing complex shapes.

2) **Application in the civil industry**

- Manufacturing experimental instruments and prototypes, and precisely manufacturing grid plates, frequency dividers, cams, electric brushes, dial scales, screens, shims, reticles, and so on;
- Often used in producing the decorative profile with all kinds of plane or shape, such as elevator doors, badges, signboards, etc;
- Forming the skeleton of an artificial heart valve, titanium alloy speed disc of a camera, resistance wire and sheets, filter screens, etc.

4.4 Basic Principle and Process

There are many complicated localized reactions in the process of chemical etching. It is not only a pure chemical process, but also may involve micro-area electrochemical reactions and some physical processes, such as diffusion and adsorption (Harris 1976).

For alloy materials, these exist simultaneously and the dominating process depends on the composition and content of metal materials. The closer the content of the high and low potential metals in an alloy is, the larger the potential difference is. In this case, the electrochemical process plays a leading role (Lim et al 2006). Otherwise, the chemical reaction has a dominating influence.

If the material is pure metal, the chemical reaction is the main process. Its mechanism is the dissolution of the grain boundaries and grains. Firstly, grain dissolution occurs, and each grain has a different dissolution rate; secondly, grain boundary dissolution also takes place. Normally the dissolution rate of the grain boundaries differs from that of the grains. Widely spaced grains are dissolved relatively quick and corrosion proceeds until an uneven surface appears. Whereas lattice distortion and enriched impurities in the grain boundary lead to the faster dissolution rate, which makes the whole grain subjected to pit corrosion. The smaller grain size brings about the lighter corrosion in the grain boundary.

There are a series of steps to be performed in a chemical etching process. The process for integral chemical etching is shown in Figure 4-5.

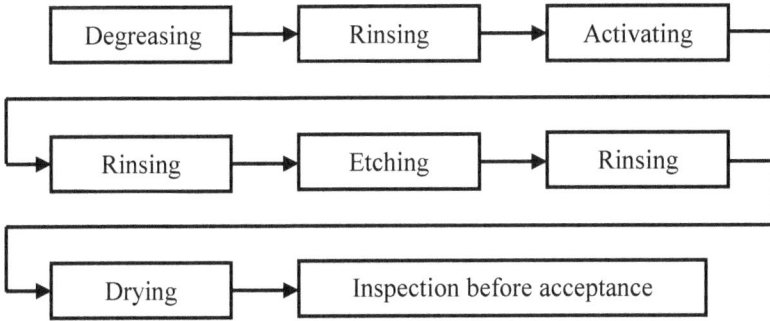

Figure 4-5 Process steps for integral chemical etching

In terms of selective chemical etching, more steps are involved. The typical process is given in Figure 4-6.

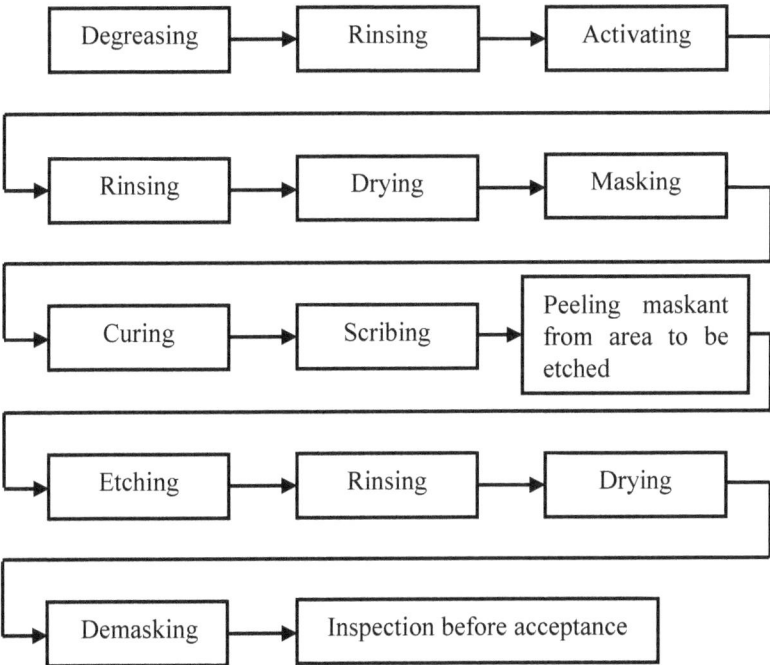

Figure 4-6 Process steps for selective chemical etching

4.4 A Brief Review of Chemical Etching for Titanium Alloys

Titanium alloys are difficult to be manufactured using conventional machining methods. This can be explained by their physical, chemical, and mechanical properties (Jin and Li 1989) as follows.

1) Poor conduction of heat

Titanium is a poor conductor of heat. Heat, generated by the cutting action, is concentrated on the small areas, and therefore does not dissipate quickly. Local high temperature causes tool wear and reduces the service life.

2) Low modulus of elasticity

Titanium has a relatively low elastic modulus, which is 1/2 of that of steel. A large elastic recovery is produced under the cutting action. The deflection tends to occur under tool pressure when cutting a deep hole in the workpiece or a longer workpiece. This causes tool rubbing and tolerance problems, even "locking" of the cutting tool.

3) Strong chemical reactivity

Titanium has a strong alloying tendency or chemical reactivity with materials of the cutting tools at tool operating temperatures. On one hand, the cutting tool material and titanium may react to produce an alloy, and on another hand titanium absorbs oxygen, nitrogen and hydrogen in atmosphere to form titanium oxide, titanium nitride and titanium hydride films. These make surface layers become hard and brittle, leading to the reduction of plasticity and the increase of work-hardening of the titanium workpiece, furthermore causing galling, welding, and smearing of the cutting tool.

Due to the above properties of titanium alloys, there are some special requirements for cutting tools and equipment. This restricts the processing quality and production efficiency of titanium parts. Particularly, it poses a challenge for precision processing of complex-shaped and thin-walled parts, and greatly limits the practical application of titanium alloys.

Chemical etching, the non-traditional processing technology, has addressed the concerns and therefore become a promising technology of shaping titanium alloy parts which cannot be easily obtained by traditional machining methods.

The steps of the chemical etching process of titanium alloy normally consist of cleaning, masking, scribing, etching and demasking (Shen 1984, AMEH 1993).

1) Cleaning

Cleaning is a vital preparatory step in the chemical etching process. It is carried out to remove the oil, grease, dust, markings, oxide layer, and any other foreign contaminants and to ensure the etching quality of the finished part. An improperly cleaned surface can result in poor adhesion of the maskant, causing a non-uniform etching rate which can lead to inaccurate final dimensions. For titanium alloys, chemical degreasing is first adopted to remove contaminants. It is performed in an alkaline solution consisting of 40 g/L sodium hydroxide, 25 g/L sodium carbonate, 40 g/L sodium phosphate, 5 g/L sodium silicate at 70-80°C for 20 minutes. Then the surface is pickled into hydrofluoric acid (5 wt.%) and nitric acid (65 wt.%) at room temperature for 3 minutes to eliminate the oxide layer and to obtain a fresh and bright metal surface.

2) Masking

Masking is an important step of coating a cleaned surface with maskant material. The maskant should not only adhere enough to the surface of the material, but also be easily strippable. Moreover, it should be inert to etchants and able to withstand heat used in the chemical etching process, which can protect the substrate and ensure that only desired areas are etched. Before masking, the workpiece must be dry. The maskant material is applied to the metal surface via 3 methods: dipping, brushing, and spraying. The selection of the method depends on the size and shape of the workpiece. Spray-masking is preferred for parts with large surface area, while dip- or brush-masking is often used in parts with small area or a complex shape. The maskant thickness is 0.3 mm. In the method of dip-masking, in order to avoid the generation of bubbles, the liquid maskant material is stirred by the mechanical method. The coated part is put into a dry oven to complete curing.

3) Scribing

Scribing is to expose the areas that are subjected to chemical etching. This is often done by hand through a scribing knife, which is guided by templates. Knifepoint must remain vertical to the part surface during scribing and the appropriate force should be used, as shown in Figure 4-7.

Knifepoint must cut completely through the maskant without scoring the metal. After scribing, the maskant in the selected area is peeled using caution to avoid damaging the remainder. This is to allow the etchant solution to dissolve the unwanted portions.

Laser scribing is gradually used to cut the maskant material nowadays, but the machine is relatively expensive.

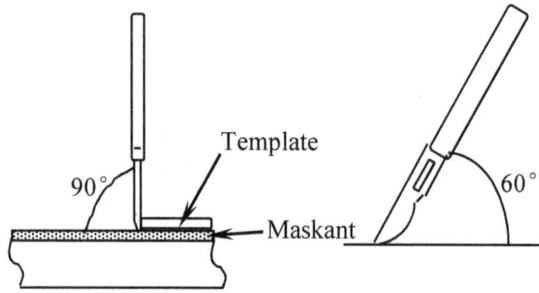

Figure 4-7 Scribing by hand through a scribing knife

4) Etching

This step is the most important stage to produce the required component from the raw part. The workpiece is immersed into the chemical etching bath and uncovered areas are etched.

The etching operation is performed in an immersion-type chemical etching device, comprising of three parts: PVC container, heating and cooling system, mechanical stirrer. The typical chemical etching set-up is given in Figure 4-8. The workpiece is hung vertically or horizontally in the PVC container. The temperature and concentration of bath constituents remained uniform by stirring.

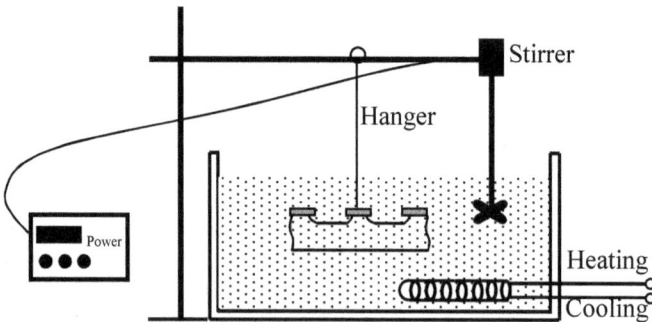

Figure 4-8 Experiment set-up for chemical etching

The etching rate is expressed as the depth of the chemical etching process per unit time. The thickness of the workpiece can be tested with 35DL ultrasonic micrometer (deviation is ±0.001) prior to and after etching. The etching rate can be calculated by Eq.(4.1):

$$v = \frac{h_0 - h_1}{t} \qquad\qquad (4.1)$$

where v is etching rate (μm/min), h_0 is original thickness of the workpiece (μm); h_1 is thickness after etching (μm); t is etching time (min).

The etching rate varies with a number of factors, including the type and concentration of bath constituents, as well as temperature conditions. In industry, due to its inconstant nature, the etching rate is often determined experimentally immediately prior to the etching process. A small sample of the material to be cut, of the same material specification, heat-treatment condition, and approximately the same thickness, is etched for a certain time; after this time, the etched depth is measured and used with the time to calculate the etching rate. For a titanium alloy workpiece, generally single sided etching is carried out due to a large amount of heat released from the etching process.

5) Demasking

After etching, the workpiece is rinsed to clean the etchant and stop further reaction. Then it is immersed into nitric acid (specific gravity 1.42 and volume ratio of less than 50%) to remove black deposits on the surface of the titanium workpiece. Further cleaning by hot water and drying are performed. The final stage is to remove the maskant material. The most common method is simple hand removal using scraping tools. The inspections of the dimensions and surface quality are accomplished before packaging the finished part.

Figure 4-9 demonstrates the major steps of chemical etching of the titanium alloy part.

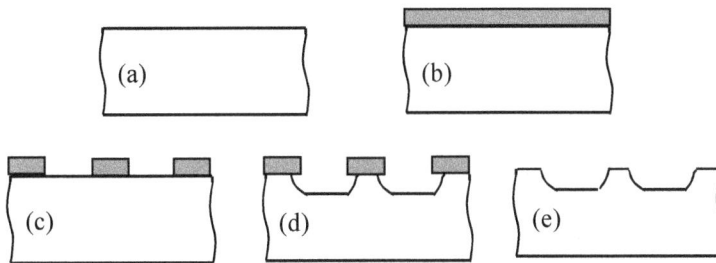

Figure 4-9 Steps of chemical etching of titanium alloy part: (a) clean raw part; (b) apply maskant; (c) scribe, cut, and peel the maskant from the areas to be etched; (d) etch; (e) remove maskant and clean to yield the finished part

The flow chart of the chemical etching process of titanium alloys is illustrated in Figure 4-10.

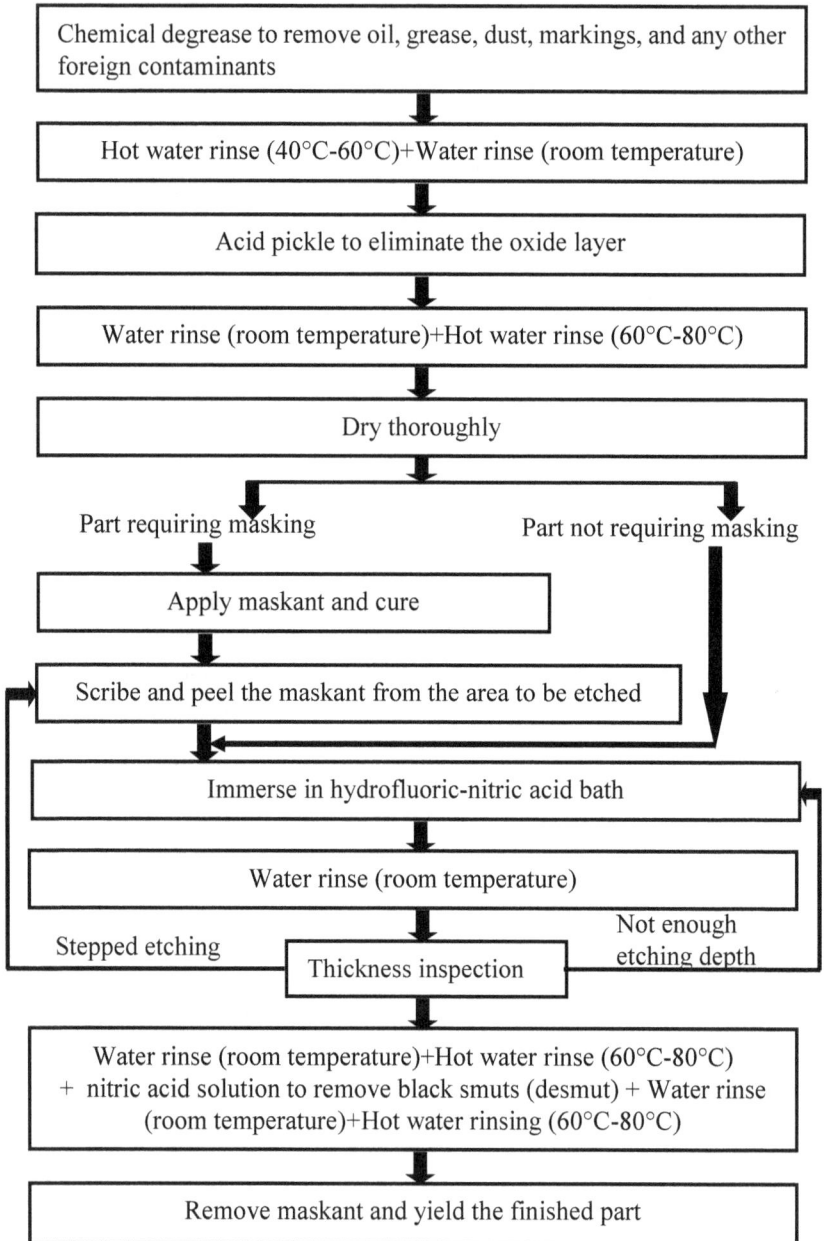

Figure 4-10 Flow chart of chemical etching process of titanium alloys

4.5 Control of Chemical Etching Process of Ti-6Al-4V

The United States, Europe and Russia began studies on chemical etching for titanium alloys in the 60s of the 20th century. In the 70s of the 20th century, Armco Steel Corporation (1973) and McDonnell Douglas Corporation (1974, 1977) did some research work on chemical etching bath of titanium alloys and patented them. In the late 80s of the 20th century, some Chinese scholars began to explore chemical etching process (Li et al 2012). From these studies, the chemical etching process for titanium alloys with more uniform metal removal rate was found (Seiji and Yukari 2006, Mogoda and Ahmad 2004, Say and Tsa 2004)

However, there are some drawbacks in the existing process: difficult control of etching rate, high surface roughness, high hydrogen absorption, negative effect on mechanical properties of titanium alloy substrate, short bath lifespan and environmental pollution. Therefore, the etching rate control, surface quality assurance and environmental pollution prevention need to be further explored. The characteristics of chemical etching depend on the type and concentration of bath constituents, as well as the temperature and etching time. To solve these problems, the research group led by Dr. Lin conducted experimental research on effective control of the chemical etching process of titanium alloys. The influencing factors and mechanisms were studied. Some of the work is presented in this section.

4.5.1 Chemical Etching Bath

Chemical etching bath is the most important component in controlling the chemical etching process of titanium alloys. The most suitable bath should have properties as follows:

- Moderate etching rate (12-18 μm/min);

- Low surface roughness on the etched workpiece (<1.6 μm);

- Low hydrogen absorption on the etched workpiece (<80 ppm);

- No effect on mechanical properties of the workpiece.

At present, the chemical etching bath of titanium alloys generally consists of a mixture of acid and surfactants.

Annealed Ti-6Al-4V sheets were used as test materials. The above properties are important criteria to measure the etching quality. Based on the influences of various factors on the above properties, the optimal process including bath constituents and process parameters can be determined (Liu 2008, Liang 2010, Lin and Hong 2011, Lin et al 2008).

4.5.1.1 Etchant

Titanium has a highly stable and cohesive oxide surface layer. In the process of chemical etching of titanium alloy, the passive film is damaged first, and then the substrate is attacked. This can be achieved by using reducing acid etchants, among which hydrofluoric acid (HF) has the strongest corrosion effect.

The concentration of hydrofluoric acid is an important parameter affecting the etching rate. Hydrofluoric acid is one of the sources of hydrogen ion. An increase in hydrofluoric acid concentration increases the free hydrogen ion and fluorine ion for attack. Figure 4-11 depicts the variation of etching rate with the concentration of hydrofluoric acid. The curve follows the exponential decay law:

$$\upsilon = 22.96 - 28.48 \exp(-m_{HF} / 76.96) \tag{4.2}$$

where υ is etching rate (μm/min), m_{HF} is hydrofluoric acid concentration (mL/L).

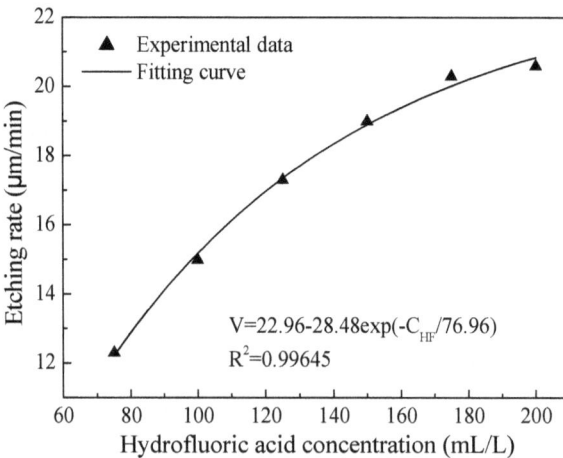

Figure 4-11 Effect of etchant concentration on etching (nitric acid 250 mL/L, temperature 30±2°C)

After a 60-minute etching period, the etching rate is 12.3 μm/min when the hydrofluoric acid concentration is 75 mL/L. When the concentration is greater than 175 mL/L, it results in a fast etching rate (more than 20 μm/min), but at the same time produces some problems, such as dishing defect, and uneven etching. The influencing extent of hydrofluoric acid on the etching rate is reduced in case of higher concentration.

But the effect of hydrofluoric acid concentration on surface roughness is minor (Figure 4-12).

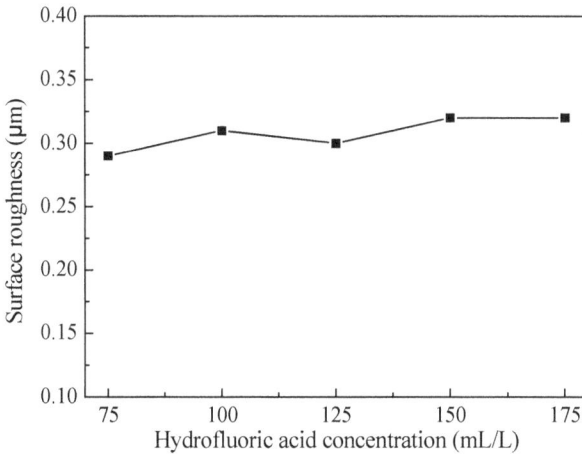

Figure 4-12 Effect of etchant concentration on surface roughness (nitric acid 250 mL/L, temperature 30±2°C)

4.5.1.2 Oxidant

Nitric acid (HNO_3) is a commonly used oxidant in chemical etching bath of titanium alloys. Hydrofluoric acid and chromic acid (H_2CrO_4) were used in early times in the United States. However, there are some disadvantages for chromic acid: high cost, environmental pollution, and hydrogen embrittlement of the etched part. Afterwards, hydrofluoric acid-nitric acid was employed. Nitric acid is a strong oxidant that can oxidize hydrogen gas to form water and inhibit hydrogen evolution, so as to reduce the absorption of hydrogen on the titanium surface.

Nitric acid has two effects. One is to supply hydrogen ion available for chemical etching, and the other is to promote the surface passivation of titanium alloys.

The influence of nitric acid concentration on the etching rate, while all other parameters remained constant, was investigated and the results are shown in Figure 4-13. The rate reaches a maximum value in the range of 2-2.5 of the volume ratio of nitric acid to hydrofluoric acid. In the condition of lower concentrations of nitric acid, the free hydrogen ion increases with nitric acid concentration, while the passive film is unstable and dissolved by acid, which indicates that the rate of corrosion is greater than that of the passive film formation. Consequently, the etching rate shows an increasing trend. However, with the further increase of nitric acid concentration, the stability of the passive film is enhanced, and the contact between bath constituents and the titanium substrate is hindered. As a result, the etching rate is reduced.

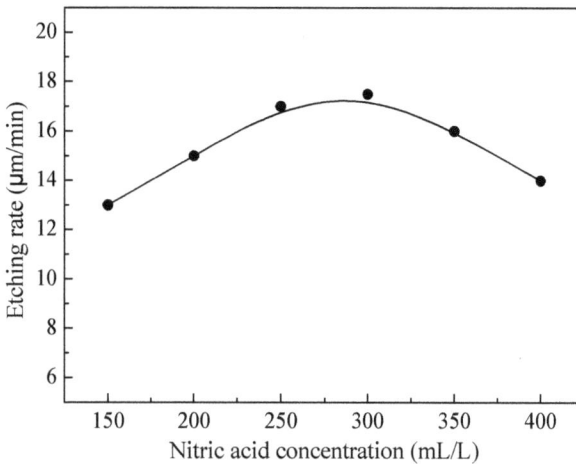

Figure 4-13 Effect of oxidant concentration on etching rate (hydrofluoric acid 125 mL/L, temperature 30±2°C)

The etched surface is dark gray and the surface roughness (Ra) is 0.42 μm in bath with only hydrofluoric acid, whereas while in bath with hydrofluoric acid and nitric acid, the surface is bright and the surface roughness is decreased to 0.30 μm. The surface morphologies of Ti-6Al-4V etched in the above two baths observed using a FEI Quanta 200 Scanning Electron Microscope are shown in Figure 4-14. The addition of nitric acid makes the recessed areas more shallow and the surface more uniform.

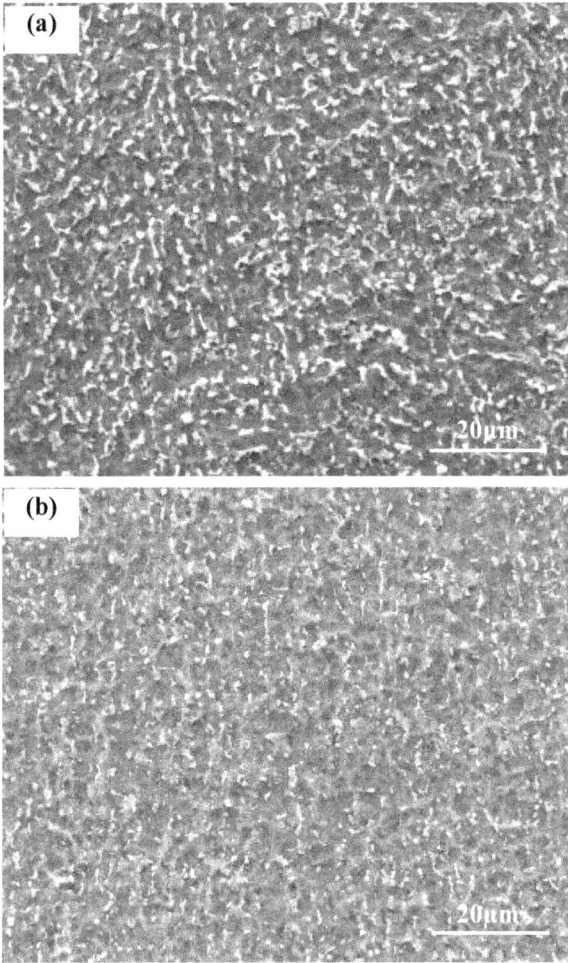

Figure 4-14 SEM images of Ti-6Al-4V in different bath: (a) hydrofluoric acid; (b) hydrofluoric acid and nitric acid

The concentration of nitric acid has a greater impact on the surface roughness. The measured data is plotted in Figure 4-15. The surface roughness presents a decreasing trend with an increase of nitric acid concentration. It can be seen from the surface quality that the higher the concentration of nitric acid is, the brighter the surface is. The surface roughness of the raw material is 0.50 μm, after etching, the surface roughness becomes 0.28 μm. Etching in hydrofluoric-nitric acid bath produces a smaller surface roughness.

During chemical etching, micro galvanic cells exist on the surface of Ti-6Al-4V. The etching rate in recessed areas is different from that in protruding areas. Corrosion products in recessed areas have difficulty diffusing into the bath solution and easily form a passive film, which slows down the rate of corrosion; on the contrary, corrosion products in protruding areas have a faster diffusion rate, and spread into bath solution in time, and passivation does not easily occur. The coactions of the fast etching rate in protruding areas and the slow etching rate in recessed areas reduce the surface roughness.

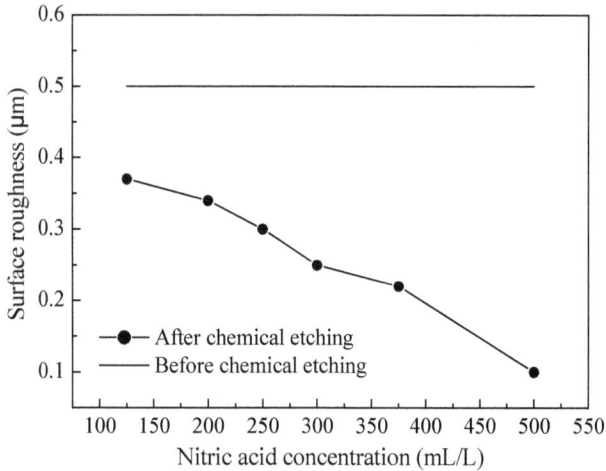

Figure 4-15 Effect of oxidant on surface roughness (hydrofluoric acid 125 mL/L, temperature 30±2°C)

From the above discussion, it can be found that the volume ratio of nitric acid to hydrofluoric acid is significant in controlling the etching quality. If the nitric acid concentration is too high, it results in a slow etching rate. On the other hand, if hydrofluoric acid concentration is too high, it results in a fast etching rate and rough surface. The optimum ratio of nitric acid to hydrofluoric acid is approximately two.

Through comprehensive consideration of the influences of hydrofluoric acid and nitric acid on the etching rate and surface roughness, the determined concentration range of hydrofluoric and nitric acid is 80-140 mL/L and 150-300 mL/L, respectively.

4.5.1.3 Surfactant

Surfactant plays an essential role in avoiding unacceptable etching defects and improving surface quality. Good surfactants have low surface tension and can wet the titanium surface. However only using one surfactant is difficult to achieve this, therefore the combined use of several kinds of surfactants is needed to meet the etching requirements.

- Uniform chemical milling rate;
- Low surface roughness on the etched workpiece;
- No effect on the function of ingredients in bath.

Besides the above requirements, the etched surface must be smooth, and without defects to the naked eye.

In bath without surfactants obvious gas channeling appears in the etching edge of Ti-6Al-4V (Figure 4-16a); when only surfactant TY is added into the bath, this phenomenon is eliminated, but the surface finish is not good (Figure 4-16b); In bath with only surfactant TN better surface finish is produced, but there are some defects, such as channelling (Figure 4-16c); If two surfactants are used at the same time, it can improve the quality of fillet, surface finish and uniform etching (Figure 4-16d).

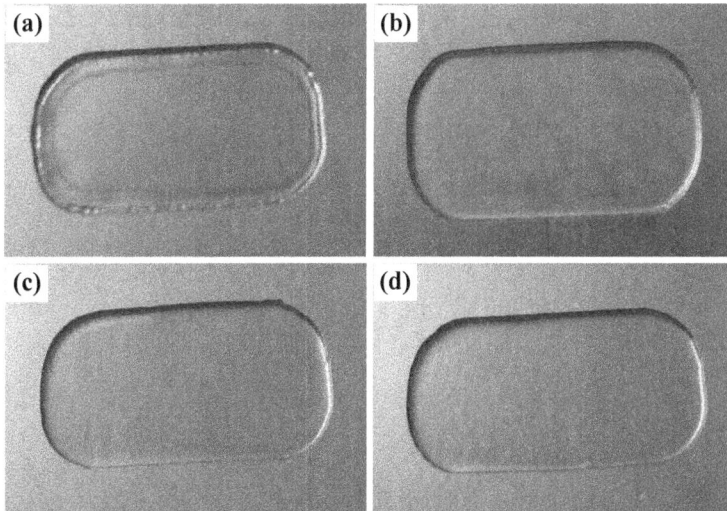

Figure 4-16 Effect of surfactants on surface quality: (a) without surfactant; (b) only TY; (c) only TN; (d) composite surfactant TY and TN

The composite surfactants orient at the bath surface and micelles form in the interface of the substrate/bath solution. They reduce surface tension and increase the contact area between the titanium surface and the bath, and then prevent the gathering of gas produced in chemical etching. These properties of surfactants result in uniform etching on the surface.

Adsorption isotherm is defined as the relationship between the amount of adsorption and the concentration of surfactant. It can reflect the degree of adsorption of surfactant on the solid surface. It can be used to obtain the maximum adsorption degree of different surfactants.

Before adsorption, the initial concentration of surfactant is C_0 and the surface tension is σ_0; after adsorption, the surface tension is σ. By plotting the curve of surface tension-concentration, the critical micelle concentration (CMC) can be found, which corresponds to the turning point in the curve. This can assist in the selection of surfactant concentration and make it work well in relatively low concentrations. The amount of absorption Γ corresponding to the concentration C can be calculated from the σ-C curve and is further used to plot the adsorption isotherm curve.

According to the Gibbs adsorption Eq.(4.3), the saturated adsorption amount of surfactant at 30°C can be obtained.

$$\Gamma = -\frac{1}{nRT} \times \frac{d\sigma}{d\ln C} \tag{4.3}$$

where Γ is the amount of adsorption, mol·m^{-2}; R is the gas constant, 8.314 J·mol^{-1}·K^{-1}; T is the absolute temperature, K; C is the concentration of surfactant, mol·L^{-1}; σ is the surface tension, N·m^{-1}.

Figure 4-17 depicts the curves of surface tension vs. surfactant concentration. It can be divided into two stages. In the first stage, the surface tension shows a decreasing tendency as the surfactant concentration increases. When the surfactant concentration reaches a

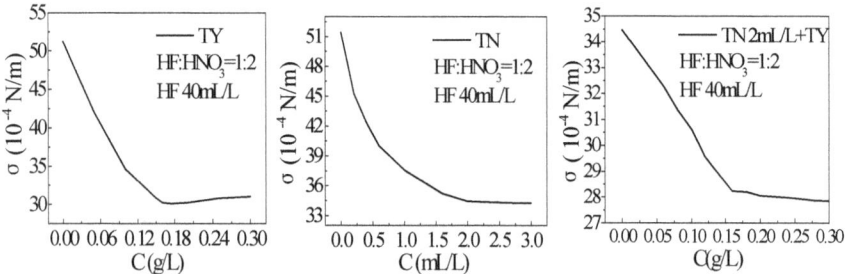

Figure 4-17 Curves of the surface tension vs. concentration of surfactants in chemical etching bath

critical value, the turning point appears in the curve, indicating it is saturated adsorption. In the second stage, the surfactant concentration absorbed on the solution surface and the surface tension remain relatively constant. As shown in Figure 4-18, at 30°C CMC is 0.16 g/L and σ_{CMC} is 30.22×10^{-4} N·m^{-1} for surfactant TY, CMC is 2 mL/L and σ_{CMC} is 34.44×10^{-4} N·m^{-1} for surfactant TN, whereas σ_{CMC} is 28.23×10^{-4} N·m^{-1} if 0.16 g/L TY and 2 mL/L TN are added to the bath to form composite surfactants, which is obviously lower than the surface tension of a single surfactant, suggesting that the composite surfactants have a better synergistic effect.

The adsorption isotherm curves of surfactants are shown in Figure 4-18. The shape of the adsorption isotherm is similar to that of the Langmuir monolayer adsorption isotherm. There are two stages for the variation of the amount of adsorption with surfactant concentration. Firstly the amount of absorption presents an approximately linear increase tendency. This is because the van der Waals forces in the molecular chains of absorbed monomer molecules lead to the production of a large number of semi micelle aggregates on the solid surface, the amount of which rapidly increases with an increase in surfactant concentration. In the next stage, the amount of adsorption becomes stable. It likely results from the desorption of surfactant monomer molecules during the collision with massive micelle aggregates produced under the surfactant concentration which is above the CMC value. The coaction of adsorption and desorption enables the adsorption of surfactants on the titanium surface to be in an equilibrium state and the amount of adsorption reaches saturation.

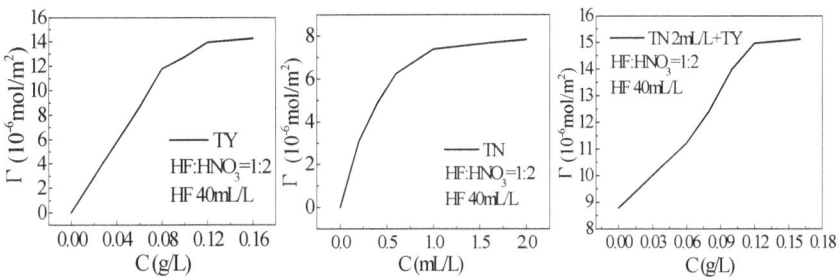

Figure 4-18 Adsorption isothermal curves of surfactants in chemical etching bath

CMC and the saturated adsorption amount of single surfactant TY, TN and composite surfactants TY and TN are listed in Table 4-1.

Table 4-1 Adsorption physical parameters of surfactants in chemical etching bath

Surfactant	$\sigma_{CMC} (10^{-4} \, N \cdot m^{-1})$	$\Gamma_{max} (10^{-6} \, mol \cdot m^{-2})$
TY	30.23	14.26
TN	34.44	7.82
TY+TN	28.23	15.10

TY plays a key role in reducing the surface tension and improving wetting. The solid surface absorbs the gas generated from the chemical dissolution reaction because of excess energy. Oriented adsorption occurs in the gas-liquid interface for surfactant TY, with the polar group toward the solution phase and the non-polar group toward the gas phase, causing the reduction of surface tension of the solution. When the TY concentration reaches a critical value in the chemical etching solution, the surface tension of the solution is lower than that of the titanium solid surface, which makes the solution spread over the solid surface, resulting in larger contact area between the chemical etching solution and the titanium surface. The original solid-gas interface is completely replaced by the solid-liquid interface. This explains the role of surfactant TY in eliminating gas channelling. TN has a good stability in the chemical etching solution and a good compatibility with TY. It increases flow and dispersion characteristics of the chemical etching solution. The gas bubbles are removed from the titanium surface in time. This improves the etching uniformity and surface brightness. After adding two kinds of surfactants (anionic surfactant TY and non-ionic surfactant TN), the surface tension is reduced to the lowest value. So the application of the composite surfactants to chemical etching bath can best improve surface properties.

The optimum concentration of TY and TN is 0.16 g/L and 2 mL/L, respectively. 0.15-0.20 g/L TY and 1.5-2.5 mL/L TN are considered to use in practical production.

4.5.1.4 Temperature

Temperature is one of the important process parameters. Proper temperature provides suitable environment for chemical etching. A great deal of heat is released due to corrosion dissolution of titanium in a chemical etching bath. Under a loading capacity (the surface area of workpiece that is etched in a certain volume of solution) of 1 dm²/20L , the temperature is increased from 28°C before etching to 36.8°C after 70 min of etching when the temperature is not controlled. The temperature rise and heat amount are plotted in Figure 4-19 and the data at different etching period is given in Table 4-2.

The amount of heat and titanium ions dissolved has a linear relationship:

$$Q = 0.27 + 26.16\, m_{Ti} \tag{4.4}$$

where m_{Ti} is titanium ion concentration in the bath, $g·L^{-1}$; Q is amount of released heat, kJ.

By formula (4.4), the heat released is 26.43 kJ when 1g/L titanium ion is dissolved. If the temperature is too high, a large amount of heat is released, leading to difficult control of the etching process. If the temperature is too low, it results in a slow etching rate and rough surface.

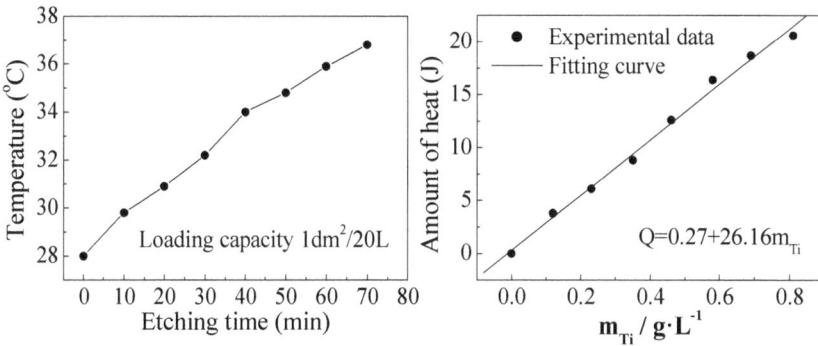

Figure 4-19 The temperature rise and amount of heat in the process of chemical etching of Ti-6Al-4V without temperature control

Table 4-2 Released heat in the process of chemical etching of Ti-6Al-4V

Time (min)	Titanium ion concentration (g/L)	Temperature rise (°C)	Amount of heat (kJ)
10	0.12	1.8	3.78
20	0.23	2.9	6.09
30	0.35	4.2	8.82
40	0.46	6.0	12.6
50	0.58	6.8	16.38
60	0.69	7.9	18.69
70	0.81	8.8	20.58

Temperature has a strong effect upon the etching rate, as illustrated in Figure 4-20. The etching rate is less than 12 μm/min when the temperature is below 298 K (25°C). As the temperature increases, the bath becomes more active, meanwhile diffusion and the chemical reaction rate are getting faster, but it is very difficult to control.

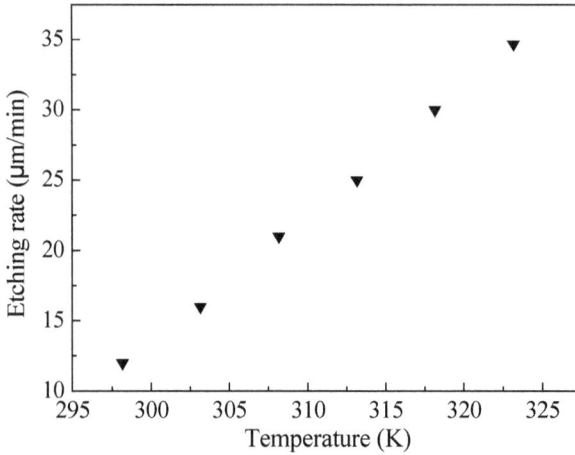

Figure 4-20 Effect of temperature on etching rate (hydrofluoric acid 125 mL/L, nitric acid 250 mL/L, TY 0.16 g/L, TN 2 mL/L)

Using the Arrhenius equation ($k=A\exp(-E_a/RT)$), the curve of the logarithm of the etching rate vs. reciprocal temperature is plotted (Figure 4-21).

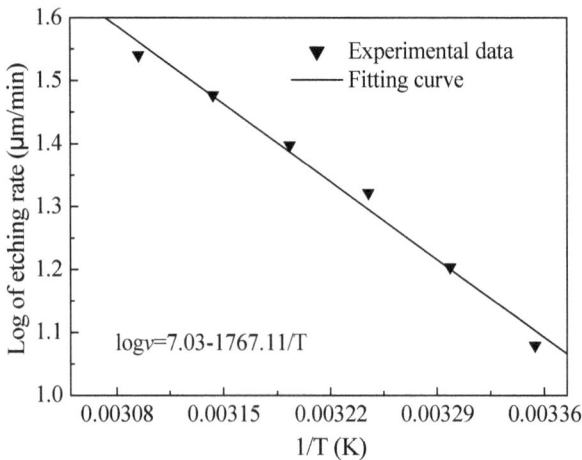

$$\log v = 7.03 - 1767.11/T$$

Figure 4-21 Logarithm of etching rate as a function of temperature

It can be described using equation (4.5):

$$\log \upsilon = C - \frac{E_a}{2.3RT} \qquad (4.5)$$

where v is the etching rate (μm/min), E_a is the activation energy, C is the constant, T is temperature (K). C and E_a can be obtained from the intercept in the Y axis and the slope ($-E_a/2.3R$), respectively.

In this work, E_a for chemical etching of Ti-6Al-4V is 33.79kJ/mol, which reveals that chemical etching in this bath is likely controlled by diffusion. The result is consistent with the observation of Figure 4-11 where the etching rate shows the tendency of slowing down in a higher hydrofluoric acid concentration.

Temperature also affects the surface roughness (Figure 4-22). With an increase of temperature, the surface roughness increases. At 45°C, the surface becomes the roughest. When the temperature is raised, corrosion and gas evolution rates increase, so that the convention and diffusion in the etching bath are accelerated. Corrosion products in recessed areas diffuse quickly and passivation of the metal surface in these areas does not occur easily. As a result, there is little difference in the etching rate between recessed and protruding areas. The low levelling effect of nitric acid leads to the increase of surface roughness. If the temperature continues to increase, corrosion dissolution further accelerates, too much products are accumulated in recessed areas and unable to completely diffuse out of these areas. Thus the levelling effect of nitric acid can function well and the surface roughness falls again.

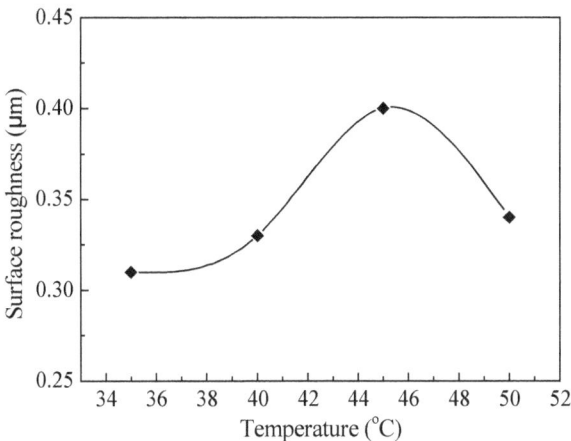

Figure 4-22 Effect of temperature on surface roughness

4.5.1.5 Stirring

Stirring is also an influential factor for etching quality. Via stirring, the diffusion process, which includes the chemical etching solution towards the titanium surface and the corrosion products on the surface of titanium towards the solution, is accelerated. In addition, the accumulation of gas bubbles on the surface is eliminated. Simultaneously the temperature of the whole etching solution is uniform, which is beneficial to uniform etching and the surface quality.

There are two ways of stirring: compressed air stirring and mechanical stirring. When mechanical stirring is adopted, the speed should be properly controlled. If the speed is too low, it easily causes uneven etching. However, if the speed is too high, the solution flows along the rotational direction of the stirrer and produces the impact on the titanium surface, which results in the generation of the swirling defect. The suitable range of mechanical stirring speed is 150-300 rpm. In chemical etching bath with a large volume, mechanical stirring is difficult to produce a good stirring effect, and therefore air stirring is used instead of mechanical stirring. Even stirring is achieved by making the compressed air discharge from multiple points in the bath.

4.5.1.6 Placement Method

In addition to stirring, how a workpiece is placed in the solution also has an impact on the etching quality. When the workpiece is hung in the horizontal direction, the accumulation and retention of gas bubbles in the edge part prevents the etching solution from corroding this part, leading to the slowdown of the etching rate and uneven etched surface (Figure 4-23a). If stirring is applied, the gathering and retention of gas bubbles are eliminated, and the uniformly-etched surface is obtained. Similarly, the phenomenon of gas retention is observed on the etched surface which is placed in the vertical direction. The gas evolves from bottom up along the etched surface and some gas bubbles get stuck in the upper part of the etched surface. Stirring can help solve this problem to some extent. But etched edges are non-uniform and the etching factor in the upper and lower sections of the part are different. Regardless of the use of stirring, gas has an upward impact on the maskant and reduces the adhesion between the maskant and the substrate. The chemical etching solution easily enters the upper section, resulting in the larger etching depth in the upper section of the part (Figure 4-23b). Table 4-3 gives the etching quality in different placement methods.

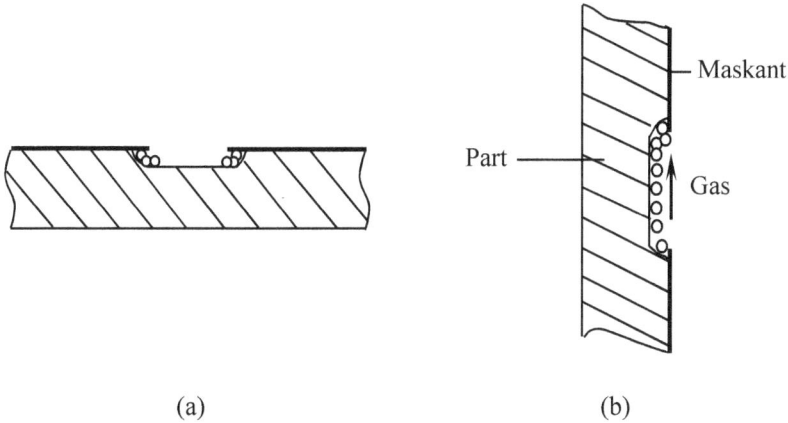

(a)   (b)

Figure 4-23 Effect of horizontal (a) and vertical (b) hanging on chemical etching

Table 4-3 Etching quality in different part placement methods

Placement method	Stirring	Surface quality
Vertical direction	No stirring	Obvious up-to-down airflow traces, non-uniform etched edge, different etching depth in the upper and lower sections of part
Vertical direction	Stirring	Unapparent airflow traces, non-uniform etched edge, different etching depth in the upper and lower sections of part
Horizontal direction	No stirring	Uneven surface
Horizontal direction	Stirring	Uniform and even surface

In pratical manufacturing most workpieces have a large size, so horizontal placement method is generally not used. Hence, the up-down etching depth difference under vertical hanging was chiefly tested. The up-down etching depth difference increases with an increase in up-down distance and etching depth. This can be improved by turning the part over. The biggest depth difference is reduced from 0.12 mm to 0.04 mm if the part is turned over every 30 minutes. The biggest depth difference is reduced from 0.12 mm to 0.02 mm if the part is turned over every 15 minutes. The up-down unevenness is greatly reduced by turning the part over (Table 4-4 and Figure 4-24).

Table 4-4 Up-down depth difference of vertical hanging part

Turnover time period (min)	Etching depth in the lower section (mm)	Etching depth in the upper section (mm)	Up-down etching depth difference (mm)
No turnover	1.12	1.0	0.12
30	1.12	1.08	0.04
15	1.12	1.10	0.02

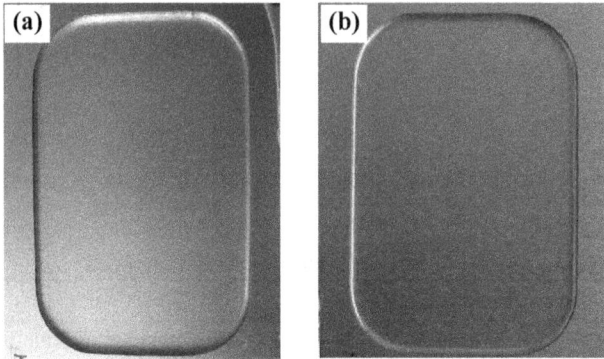

Figure 4-24 Effect of turnover on surface evenness for Ti-6Al-4V (a) without turnover; (b) turnover every 15 minutes

4.5.1.7 Titanium Ion

As the chemical etching process continues, the concentration of titanium ion in bath is increased, which will have an effect on the etching rate and quality.

With an increase in the dissolved titanium ion concentration, the etching rate is reduced. When the titanium ion concentration is greater than 30 g/L, the etching rate is less than 8 μm/min. Table 4-5 lists the results of the etching rate, surface roughness and evenness at different concentrations of titanium ion in bath. In general, the following can be observed (Figure 4-25).

1) At low concentrations of titanium ion (0-15 g/L), the concentration of hydrofluoric acid is high, resulting in a fast etching rate;

2) When the titanium ion ranges from 15 g/L to 23 g/L, the etching rate slows down;

3) When the titanium ion concentration is higher than 30 g/L, the surface roughness is increased, and the etching rate and evenness are decreased. The rise of the titanium ion concentration causes an increase in solution viscosity. Simultaneously, the decrease of the

concentrations of nitric acid and surfactants lead to a drop in wettability. Channeling and ridging defects occur;
4) If the titanium ion concentration reaches 40 g/L, the etching rate falls to 2.5 μm/min. There is a sharp increase of the surface roughness owing to the low concentration of nitric acid. However, surface evenness shows some improvement.

Table 4-5 Effect of titanium ion concentration in bath on chemical etching process

Titanium ion concentration (g/L)	Etching rate (μm/min)	Surface roughness (μm)		Depth difference (mm)
		Transverse	Longitudinal	
0	17.8	0.34	0.20	0.02
15	12.4	0.26	0.19	0.00
23	9.3	0.26	0.20	0.08
30	6.3	0.30	0.20	0.10
35	4.0	0.40	0.28	0.30
40	2.5	0.47	0.40	0.10
42	2.2	0.74	0.64	0.04

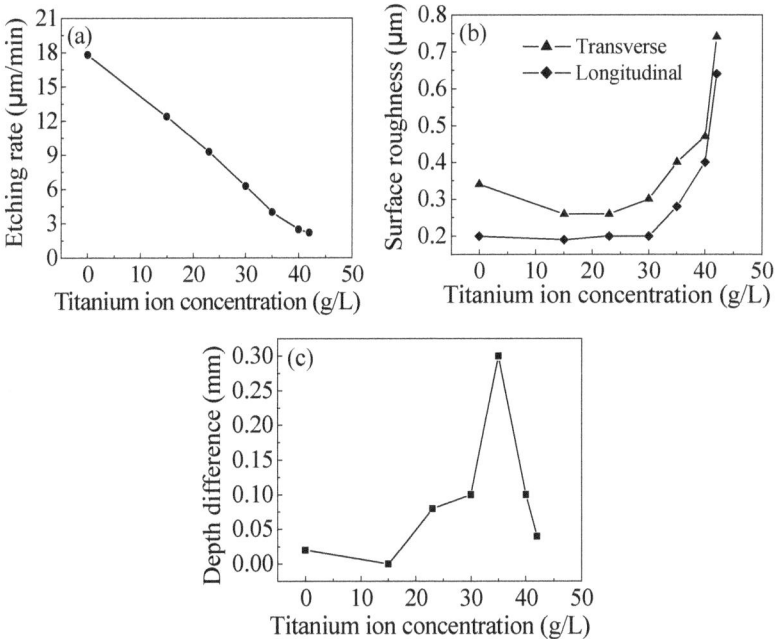

Figure 4-25 Effect of dissolved titaniunm ion concentration on (a) etching rate; (b) surface roughness; (c) depth difference

4.5.1.8 Summary

This section provides a summary of the main factors responsible for the etching performace.

1) Etching rate

Etching rate is one of the most important properties in chemical etching. The ideal etching rate for a titanium alloy is controlled in the range of 10-20 μm/min/single side. The main influencial factors are listed in Table 4-6. Higher hydrofluoric acid, lower nitric acid, higher temperature and lower dissolved titanium ion concentrations provide a higher etching rate.

Table 4-6 The main influential parameters of etching rate

Factors	Result
Hydrofluoric acid concentration	The higher the hydrofluoric acid concentration is, the faster the etching rate. (Figure 4-11)
Volume ratio of nitric acid to hydrofluoric acid	With an increase in volume ratio, the etching rate increases. In the volume ratio of 2:1, the etching rate reaches the maximum value. After that, etching rate declines. (Figure 4-13)
Temperature	Higher temperature promotes chemical etching. There is an increase of 10 μm/min if the temperature is increased by 10°C. (Figure 4-20)
Dissovled titanium ion concentration	There is a drop in etching rate as an increase of dissolved titianium ion concentration in bath. (Figure 4-25a)

2) Surface roughness

Surface roughness is another factor to be considered. The main influential factors of surface roughness are given in Table 4-7. It can be seen from Table 4-7, the surface roughness is mainly affected by nitric acid concentration, temperature and dissolved titanium ion concentration.

3) Surface quality

The etched surface must be uniform and is essentially free from defects, such as channelling, islands, gas channelling, overhang, dishing and ridging. The main influential factors include surfactant, dissolved titanium ion concentration and stirring, as shown in Table 4-8.

Table 4-7 The main influential factors of surface roughness

Factors	Result
Surface state of raw material	Good surface state of raw material results in low surface roughness of the etched area.
Nitric acid concentration	Surface roughness significantly decreases when nitric acid concentration is increased. (Figure 4-15)
Temperature	Surface roughness increases with an increase of temperature. At 45°C it reaches the highest point, and then drops. (Figure 4-22)
Dissovled titanium ion concentration	Surface roughness decreases steadily firstly, then increases rapidly, with continuous carrying out of chemical etching. (Figure 4-25b)

Table 4-8 The main influential factors of surface quality

Factors	Result
Surfactant	The use of composite surfactants effectively reduces surface tension, increases wettability and removes gas under maskant, therefore improving etching uniformity and eliminating etching defects. (Figure 4-16)
Dissovled titanium ion concentration	Some defects occur in the higher concentration of titanium ion.
Stirring	Ridging occurs in the edge of the etched surface without stirring. (Table 4-3)

Figure 4-26 shows Ti-6Al-4V specimens etched using the bath formula obtained in this research work. The bath contained hydrofluoric acid (125 mL/L), nitric acid (250 mL/L), TY (0.16 g/L), TN (2 mL/L). The chemical etching was performed at 30±2°C. The size of specimens was 120mm×90mm×2mm. The etched depth was 0.8mm.

4.5.2 Substrate Properties after Chemical Etching

Besides etching performance, the titanium substrate properties after chemical etching should be checked to optimize the chemical etching process below (Hu 2011, Lin et al 2015b).

4.5.2.1 Hydrogen Absorption

Hydrogen is easily absorbed into titanium because hydrogen has a great affinity with titanium. As stated in section 2.4, the hydrogen content in a titanium alloy should not be higher than 80-150 ppm, otherwise hydride forms, resulting in the occurrence of hydrogen embrittlement and the decrease of the fatigue strength of the titanium substrate.

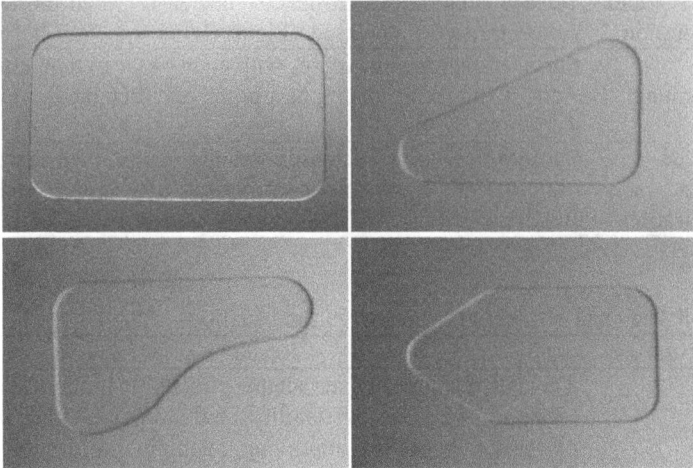

Figure 4-26 Appearance of Ti-6Al-4V specimens etched using the bath formula obtained in this research work

The difference of hydrogen pickup for different types of titanium alloy is associated with the phase composition in the microstructure. α and β phases are hcp and bcc structure, respectively. From the binary phase diagram of Ti-H system, the solubility of hydrogen in β-Ti is far greater than that in α-Ti. The maximum solid solubility of the former is 2%, the latter is only 0.19%. Therefore α-β titanium alloys are more sensitive to hydrogen than α titanium alloys (Takasaki 1994). The hydrogen content of Ti-6Al-4V before chemical etching is 56.855 ppm, and after chemical etching it is 50-60ppm, which is lower than 150 ppm.

The hydrogen content before and after chemical etching can be determined by high frequency heating gas chromatography. The hydrogen content fluctuates with the etching depth, as shown in Figure 4-27. In the beginning of etching, the amount of hydrogen diffusing into the substrate is not sufficient to compensate the amount of hydrogen going into the solution in the process of titanium dissolution. The rate of reaction is greater than the rate of diffusion, resulting in a decline in the hydrogen content in the substrate. As the etching depth increases, the propagation rate of the hydrogen absorption layer is higher than the etching rate, hence the hydrogen content in the substrate goes up again. When the etching depth is 0.9 mm, the total hydrogen content is less than 65 ppm and the difference between the average value before and after chemical etching does not exceed 20 ppm. There is no need for hydrogen-eliminating treatment after chemical etching.

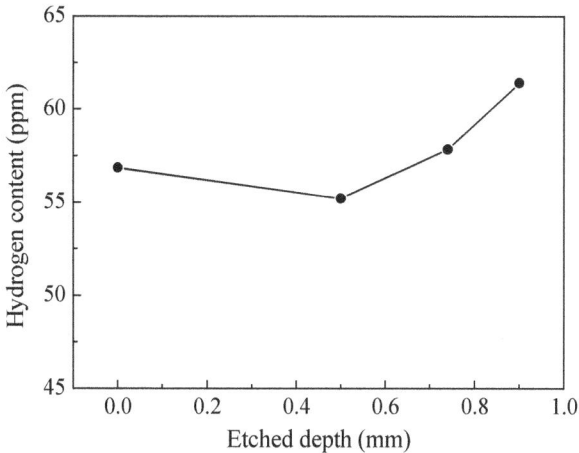

Figure 4-27 Hydrogen content as a function of etched depth in the process of chemical milling of Ti-6Al-4V

4.5.2.2 Mechanical Properties

In engineering application, the mechanical properties are the main basis for structure design under static loading. The influence of chemical etching on the mechanical properties of the substrate material at ambient temperature varies with the material and its state. In order to have an understanding of the effect of chemical etching on the mechanical properties of Ti-6Al-4V, load-displacement curves of chemical etching, mechanical machining and raw Ti-6Al-4V specimen were tested by a Material Testing Machine. The specimens were prepared according to ASTM E8 (standard test methods for tension test of metalic material). From the load-displacement curve, the parameters, such as σ_b (tensile strength), σ_s (yield strength), δ_5 (elongation), and ψ (reduction of area) can be obtained (Lin et al 2015b).

Figure 4-28 shows load-displacement curves of Ti-6Al-4V before processing, after chemical etching and after mechanical machining. Table 4-9 lists the mechanical performance indices.

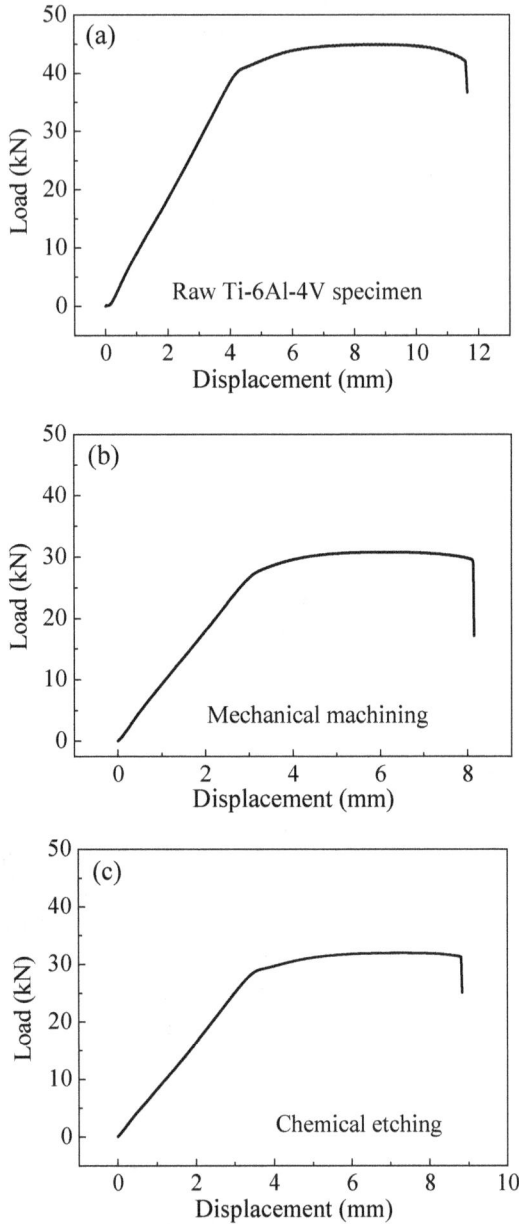

Figure 4-28 Load-displacement curves of Ti-6Al-4V before processing, after mechanical machining and after chemical etching

Compared with the raw specimen, the tensile strength and yield strength after chemical etching is decreased by 4.5% and 3.5%, respectively. Whereas in contrast with mechanical machining, the tensile strength and yield strength after chemical etching is decreased by 0.7% and increased by 0.7%, respectively. Overall, there is some reduction in the tensile strength and plasticity of Ti-6Al-4V after chemical etching. But the tensile strength after chemical etching is similar to that of after mechanical machining.

Table 4-9 Mechanical properties of raw Ti-6Al-4V specinen and Ti-6Al-4V processed using different methods

Processing method	Tensile strength σ_b (MPa)	Yield strength $\sigma_{0.2}$ (MPa)	Elongation δ_5 (%)	Reduction of area ψ (%)
Chemical etching	993	906	9.3	20.0
Mechanical machining	1000	900	9.6	21.6
Raw specimen	1040	939	11.9	23.0

4.5.2.3 Fatigue Property

The parts or components in aerospace vehicles sometimes bear high alternating stress, for example in the taking-off and landing of an airplane. It is different from high altitude flying. This alternating stress can even exceed the yield strength of a material. The frequency of loading is also very low and a certain amount of plastic deformation in each cycle takes place, resulting in a low fatigue life of these parts. Chemical etching is a processing method for parts of the aerospace and aviation industries, so the influence of chemical etching on the fatigue life of titanium alloys needs to be considered (Lin et al 2015) . Due to large variance in the fatigue test data, at least three specimens need to be measured under the same stress level. Based on the fracture situation and data dispersion degree, the number of test specimens may need to be increased.

The fatigue property after chemical etching is compared with that after mechanical machining. In the fatigue test, the frequency is 50-60 Hz, the stress ratio R (ratio of minimum stress to maximum stress in one cycle of loading in a fatigue test) is 0.1.

The fatigue test data is given in Table 4-10. The corresponding fatigue curve is depicted in Figure 4-29.

Table 4-10 Fatigue data of specimens after chemical etching and mechanical machining

Chemical etching

Specimens numbers	Applied stress (MPa)	No. of cycles
4	650	4.1894×10^4; 5.1989×10^4; 7.0486×10^4; 4.5412×10^4
3	600	8.0051×10^4; 6.9326×10^4; 5.6208×10^4
6	550	9.159762×10^6; 8.3731×10^4; 1.27480×10^5; 5.8100×10^4; 8.227862×10^6; 9.6372×10^4
6	500	1.112742×10^6; 7.9081×10^4; 1.97386×10^5; 4.880532×10^6; 10^7; 10^7
3	450	10^7; 10^7; 10^7 unbroken

Mechanical machining

Specimen numbers	Applied stress (MPa)	No. of cycles
4	650	4.0845×10^4; 2.4809×10^4; 2.3922×10^4; 3.6652×10^4
4	600	2.8865×10^4; 6.0247×10^4; 3.9806×10^4; 4.3368×10^4
4	550	1.25157×10^5; 1.13564×10^5; 6.3459×10^4; 8.4584×10^4
3	500	1.24191×10^5; 1.09543×10^5; 1.25955×10^5
7	450	7.1008×10^4; 1.33493×10^5; 7.4031×10^4; 1.29793×10^5; 1.06869×10^5; 10^7; 10^7
3	400	10^7; 10^7; 10^7 unbroken

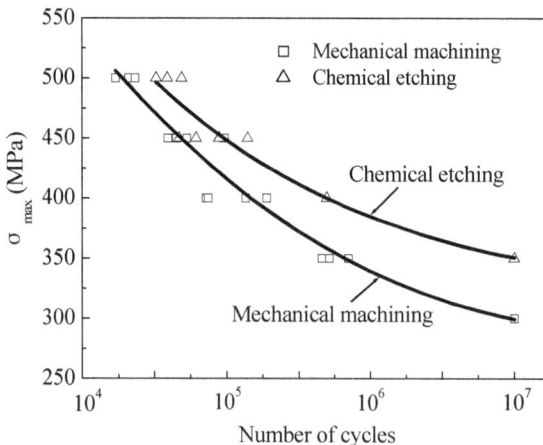

Figure 4-29 Fatigue curves of Ti-6Al-4V after chemical etching and mechanical machining

The fatigue curve of the chemical etched specimen is located above that of the mechanical machined specimen. Under the same stress level, the cycle number of the chemical etched specimen is larger than that of the mechanical machined specimen. The corresponding stress of the fatigue limit of the chemical etched specimen is higher than that of the mechanical machined specimen. When the number of cycles is 10^7, the corresponding stress of the chemical etched specimen and the mechanical machined specimen is 450 MPa and 400 MPa, respectively. Therefore, Ti-6Al-4V after chemical etching possesses a better fatigue property than that after mechanical machining. This may result from the difference in profile of the processed edge. The chemically etched specimen has a circular arc edge, which is beneficial to the fatigue property, while mechanically machined specimen has a sharp corner in the edge, which is detrimental to the fatigue property.

4.5.3 Discussion of Mechanism of Chemical Etching

The discussion of corrosion dissolution of chemical etching (Lin et al 2010) and the etching kinetics law (Lin et al 2016a) are included in this section.

4.5.3.1 Open Circuit Potential

1) In solution with only hydrofluoric acid

Figure 4-30 shows OCP curves of Ti-6Al-4V in solution containing only hydrofluoric acid. With an increase of the concentration of hydrofluoric acid, the OCP moves towards negative direction. During the initial tens of seconds, the OCP in three cases all shows a rapid negative shift, and then maintains at a relatively stable value. The value remains constant of -0.99~-1.00 V in the solution with 75 mL/L, 100mL/L and 125 mL/L hydrofluoric acid, respectively. The titanium surface in the air is usually covered with a thin layer of oxide. When Ti-6Al-4V is submerged in the solution, the surface oxide film is dissolved by hydrofluoric acid, causing the rapid negative shift of OCP. After the film is completely dissolved, hydrofluoric acid is in direct contact with the titanium substrate. With the dissolution process continuing, the OCP slowly tends to even out. After etching, the electrode surfaces are grayish black and show serious corrosion.

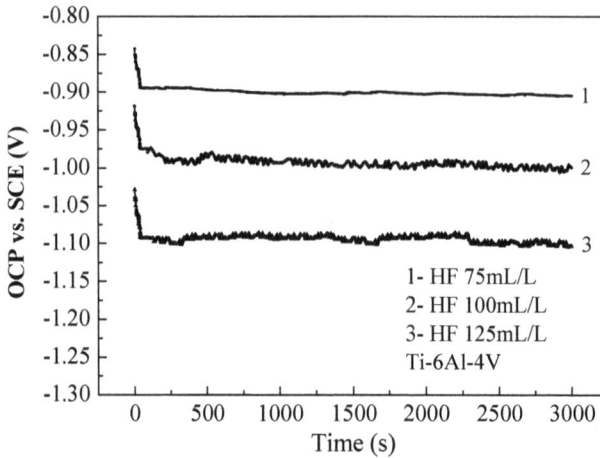

Figure 4-30 Open circuit potential versus time of Ti-6Al-4V in solution with hydrofluoric acid

The passive layer on the surface of Ti-6Al-4V is destroyed in solutions containing only hydrofluoric acid and where the fresh substrate is exposed. Ti is oxidized to Ti^{3+} or reacts with F^- to form a complex compound, simultaneously H_2 gas is produced in the cathode.

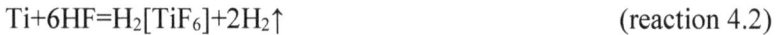

$$2Ti+6HF=2TiF_3+3H_2\uparrow \qquad\qquad\qquad (reaction\ 4.1)$$

$$Ti+6HF=H_2[TiF_6]+2H_2\uparrow \qquad\qquad\qquad (reaction\ 4.2)$$

2) In solution with only nitric acid

Figure 4-31 shows OCP curves of Ti-6Al-4V in solution containing only nitric acid.

The range of OCP is -0.3~-0.1 V, -0.01~0.07 V and -0.01~0.31 V in the solution with 75 mL/L, 100 mL/L and 125 mL/L nitric acid, respectively. It is more positive in comparison with when in solution with only hydrofluoric acid. Nitric acid is able to passivate the surface of Ti-6Al-4V, hence no corrosion occurs. The specimen has a bright surface after taken out from solution.

3) In solution with hydrofluoric-nitric acid

Figure 4-32 shows OCP curves of Ti-6Al-4V in the solution containing hydrofluoric-nitric acid. The concentration of hydrofluoric acid is 100 mL/L and the volume ratio of nitric acid to hydrofluoric acid is 1:1, 2:1, 3:1 and 4:1. The variation tendency of OCP with time in all cases has

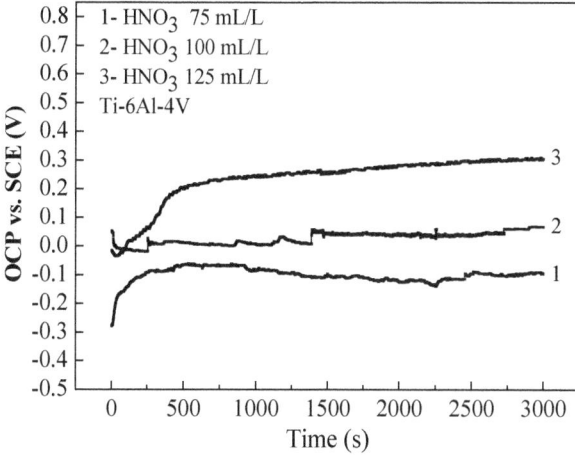

Figure 4-31 Open circuit potential versus time of Ti-6Al-4V in solution with nitric acid

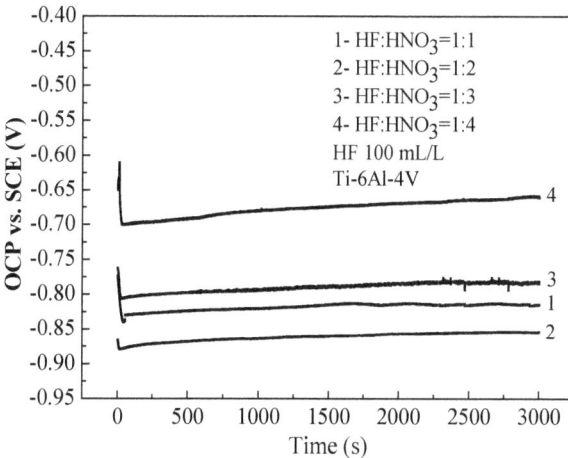

Figure 4-32 Open circuit potential versus time of Ti-6Al-4V in solution with hydrofluoric-nitric acid

a similar pattern. In the initial 60 s, the OCP moves towards negative direction sharply, then it shifts in positive direction slowly. Eventually it levels off. This can be explained as follows: the passive film on the surface can be quickly dissolved and destroyed when Ti-6Al-4V is immersed into solution, and the solution is in direct contact with the substrate. So in the beginning the phenomenon of the negative shift of the OCP is observed. With the continuation of chemical etching, the surface passive film repairs due to the presence of nitric acid, hence the OCP shows a positive shift. There are two concurrent processes, the growth of passive film and the destruction of passive film. After a period of time, the OCP tends to remain level. This is because the growth rate of passive film is equal to the dissolution rate of the titanium substratete.

The stable OCP is -0.82 V in the case of 1:1 volume ratio of nitric acid to hydrofluoric acid. If the volume ratio is increased to 2:1, the stable OCP shifts 0.04 V towards the negative direction. When the volume ratio is 3:1, the OCP is stable in -0.79 V. If the volume ratio is further increased to 4:1, the OCP is stable in -0.66 V. In hydrofluoric-nitric acid solution with the volume ration of 2:1, the stable OCP is the most negative.

4.5.3.2 Potentiodynamic Polarization Curve

The potentiodynamic polarization test is further used to study the electrochemical behavior of chemical etching in hydrofluoric-nitric acid and the effect of the volume ratio of nitric acid to hydrofluoric acid. Hydrofluoric acid concentration is 100 mL/L. The volume ratio of nitric acid to hydrofluoric acid is 1:1, 2:1, 3:1 and 4:1. The logarithm of the current is plotted as a function of the applied potential, as shown in Figure 4-33. Table 4-11 gives the electrochemical parameters obtained from the polarization curves. Passive characteristics can be seen in the anodic polarizaiton region, indicatinng the growth of passive film on the surface of Ti-6Al-4V. With an increase in the volume ratio, the anodic curves vary in regular pattern. Figure 4-33a, b, c exhibit the transition from active to passive state. The volume ratio is changed from 1:1 to 2:1, the maintaining passive current density becomes larger and the corrosion potential has a negative shift. With a further increase of the volume ratio, the maintaining passice current density decreases and the corrosion potential has a positive shift. The polarization curve shows a self-passivation trend when the volume ratio is 4:1, in this case passive film rapidly grows. The potentiodynamic polarization test results are in accordance with the results of OCP. Accordingly, the etching rate is determined by hydrofluoric acid concentration, the volume ratio of nitric acid to hydrofluoric acid.

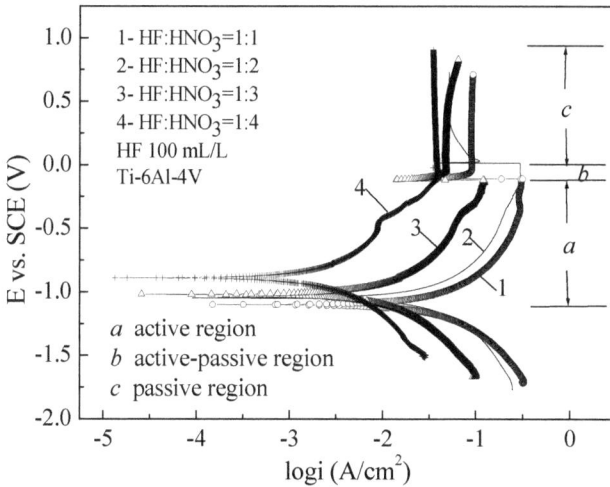

Figure 4-33 Polarization curves of Ti-6Al-4V in solution with different voume ratio of nitric acid to hydrofluoric acid

Table 4-11 Electrochemical parameters acquired from polarization curves

$V_{HNO3}: V_{HF}$	1:1	2:1	3:1	4:1
E_{corr} (V)	-0.8994	-1.0967	-1.0233	-0.8788
E_p (V)	0.0411	-0.0415	-0.1218	-0.1585
i_p (A·cm^{-2})	0.0769	0.0891	0.0440	0.0346

4.5.3.3 Kinetics of Chemical Etching

There are two processes, anodic dissolution and cathodic gas evolution, which simultaneously exist in the chemical etching of titanium alloys. The electrons that are lost in the anodic oxidation reaction are consumed by the cathodic reduction reaction, along with the gas evolution. Therefore, the overall reaction process is made up of a pair of conjugate reactions, that is, $v_{anode} = v_{cathode}$.

The chemical etching process for titanium alloys can be described as the following steps:

a) Hydrofluoric acid damages the passive film on the titanium surface and the fresh substrate is exposed.

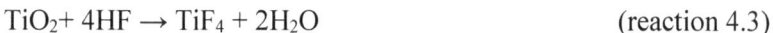

$$TiO_2 + 4HF \rightarrow TiF_4 + 2H_2O \qquad \text{(reaction 4.3)}$$

b) The substrate reacts with hydrofluoric acid to form a titanium ion.

$$Ti + 4HF \rightarrow TiF_4 + 2H_2\uparrow \qquad \text{(reaction 4.4)}$$

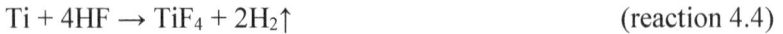

c) Hydrogen gas is oxidized by nitric acid.

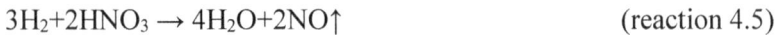

$$3H_2+2HNO_3 \rightarrow 4H_2O+2NO\uparrow \qquad \text{(reaction 4.5)}$$

d) Nitric acid contacts the substrate surface and $TiO(NO_3)_2$ passive film is produced.

$$3Ti+10HNO_3 \rightarrow 3TiO(NO_3)_2 +5H_2O+4NO\uparrow \qquad \text{(reaction 4.6)}$$

e) Hydrofluoric acid dissolves the passive film.

$$TiO(NO_3)_2+4HF=TiF_4+2HNO_3+H_2O \qquad \text{(reaction 4.7)}$$

f) The process of "attack $\rightarrow$ oxidation $\rightarrow$ dissolution" repeats continuously, which contributes to the chemical etching.

The overall chemical reaction is written as follows:

$$3Ti + 12HF + 4HNO_3 = 3TiF_4 + 8H_2O + 4NO\uparrow \qquad \text{(reaction 4.8)}$$

To investigate the dissolution behavior of chemical etching, the anodic dissolution rate is measured when the volume ratio of nitric acid to hydrofluoric acid is 2:1 and 3:1 in the range of 0 to 3000 seconds and the results are illustrated in Figure 4-34. In the beginning, the anodic dissolution rate rises quickly, and then there is a slight decline. After 300 seconds, the anodic dissolution rate becomes relatively constant.

From the results, the following can be observed:

- Anodic dissolution rate is increased by 4.2-5.1 μm/min when the solution temperature is increased by 5°C (see Table 4-12).

- Anodic dissolution rate is decreased by 0.7-1.7 μm/min when the volume ratio of nitric acid to hydrofluoric acid is 3:1 in contrast with when the volume ratio of nitric acid to hydrofluoric acid is 2:1. An increase of the nitric acid concentration improves the passivation ability and enhances the corrosion resistance of the passive film, preventing the dissolution of titanium substrate to some extent. Consequently, there is a drop in the anodic dissolution rate.

- The curve of the anodic dissolution rate variation has a similar pattern to the OCP curve.

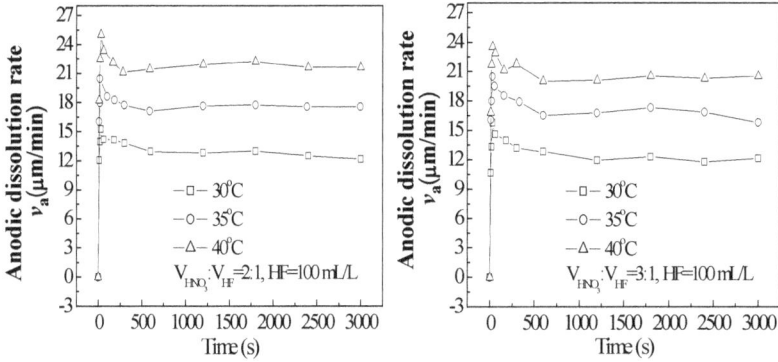

Figure 4-34 The variation of anodic dissolution rate with time of Ti-6Al-4V in etching solution

Table 4-12 Anodic dissolution rate of Ti-6Al-4V in etching bath

Temperature (°C)	The volume ratio of HNO$_3$ to HF	v_a (μm/min)
30	2:1	13.0
	3:1	12.3
35	2:1	18.1
	3:1	17.4
40	2:1	22.3
	3:1	20.6

The results of produced gas flow rate when the volume ratio of nitric acid to hydrofluoric acid is 2:1 and 3:1 are shown in Figure 4-35 and Table 4-13. The variation trend is similar to the anodic dissolution rate curve. Initially, the gas production rate rises rapidly, then it slows down, and finally stabilizes. The gas production rate is increased from 3.65 mL/min·cm^{-2} to 5.30 mL/min·cm^{-2} if the temperature is changed from 30 to 40°C. The rise of nitric acid concentration can hinder the dissolution reaction, leading to the decrease of gas evolution.

The theoretical value of the produced gas can be calculated based on the overall reaction (4.8). Assume that B is the ratio of the theoretical value of the produced gas to the total gas amount collected from the experiment. The B value is in the range of 1.13 to 1.19 when the volume ratio of nitric acid to hydrofluoric acid is 2:1, and the B value is in the range of 1.03 to 1.07 when the volume ratio is 3:1. It can be inferred from the B value that a small amount of hydrogen gas is generated in the chemical etching process. The B value exhibits a drop in the higher concentration of nitric acid. This is because the strong oxidation of nitric acid makes most of the free hydrogen ions participate in the cathodic reaction to form nitrogen oxides. Hydrogen gas is formed in the secondary cathodic reaction, which explains why the B value is greater than 1 and the range is from 1 to 1.2. This is consistent with the theoretical calculation. This is also the reason why nitric acid can reduce the hydrogen absorption of the titanium substrate.

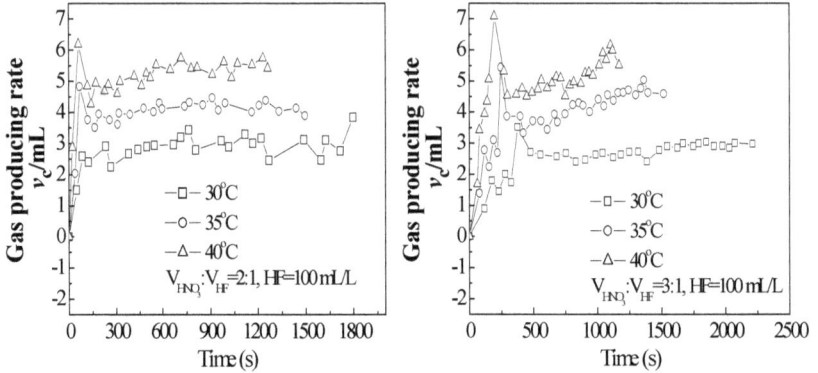

Figure 4-35 The variation of cathodic gas production rate with time of Ti-6Al-4V in etching solution

Table 4-13 Cathodic gas production rate of Ti-6Al-4V in etching solution

The volume ratio of HNO₃ to HF	Temperature (°C)	v_c (mL/min·cm⁻²)	B
2:1	30	3.65	1.19
	35	4.29	1.13
	40	5.30	1.15
3:1	30	3.32	1.03
	35	4.01	1.09
	40	4.90	1.07

From the results of OCP, potentiodynamic polarization and kinetics, the chemical etching behavior can be summarized as below:

The etching rate is chiefly determined by the volume ratio of nitric acid to hydrofluoric acid. In the condition of 2:1 volume ratio, the etching rate is the greatest.

At the initial period of chemical etching, the fluorine ion produced through the dissociation of hydrofluoric acid penetrates the oxide film naturally formed on the surface of titanium alloys, and parts of them occupy the vacancies of oxygen ions, meanwhile fluorine interstice forms.

Eventually a soluble TiF_4 layer generates and substitutes the protective oxide layer by the reaction (4.3) mentioned above. When the fresh titanium substrate is exposed in the solution, it is corroded by the hydrogen ion, hence the OCP moves toward negative direction and a faster etching rate occurs. With the extension of time, nitric acid promotes the surface passivation, which results in an decrease of the etching rate. The growth and destruction of the passive film simultaneously occur. The OCP and etching rate tend to be stable when the growth of the passive film and dissolution of the substrate achieve a dynamic balance.

4.6 Environmental Concerns

Environmental issues in chemical etching operations may be the most important factor that affects whether the process should be used or not.

1) Acid fog suppression

The bath for titanium alloys is an acidic corrosion solution, and the concentrations of hydrofluoric and nitric acid are high. During etching, acid fog is created and pollutes the environment, which limits the development of chemical etching of titanium alloys. Therefore, control of acid fog and reduction of environmental pollution are the problems to be solved. In addition to improve the quality of the etched surface, the surfactants also can suppress the acid fog by means of foam seal. When the bath reacts with the workpiece, surfactant molecules arrange with orientation on the surface of the bath (Figure 4-36), and then a layer of white foam densely covers the surface of the bath (Figure 4-37). It blocks the spread of the gas, and inhibits the volatilization of acid fog. Moreover, it reduces the loss of the effective constituents of the bath. The surfactants should be stable, and should not participate in any chemical reactions.

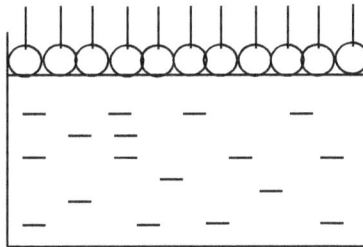

Figure 4-36 The orientation of surfactant molecules on the surface of the etching bath

Figure 4-37 A layer of white foam on the surface of the etching bath

2) Regeneration of waste chemical etching bath

In a waste chemical etching bath of titanium alloys, there are hydrofluoric and nitric acids in excess amounts, as well as a large number of dissolved titanium ions. These acids are very corrosive and are considered "hazardous waste". Because titanium is a heavy metal, there is a very strict emission limit value. However, handling and disposal of them are very costly. Therefore, regeneration of waste bath of titanium alloys becomes very important. From the perspective of "remanufacturing, recycle and reuse", the regeneration processes contain the removal of dissolved titanium ions from the waste bath and the addition of consumed constituents to the waste bath. The purpose is to recover the activity of the waste bath and make it reusable. Hence, it can extend the life span of the chemical etching bath and improve its utilization. Most importantly, it can reduce the discharge of the waste bath.

In HF-HNO_3 chemical etching bath, titanium ions can be removed from the waste bath by use of sodium or potassium salt to generate potassium fluotitanate or fluorine sodium titanate, which is insoluble or slightly soluble in acidic solution at low temperatures.

After the removal of titanium ions, HF and HNO_3 are added to supplement the consumed amount via analysis. In terms of surfactants, experiments and experience are needed to determine how much will be added as supplement. The ideal situation is to return it back to its starting form. More work is suggested to explore the recovery of titanium.

Acid fog suppression and regeneration of the waste bath can minimize or eliminate environmental pollution from occurring during and after the chemical etching process of titanium alloys.

Chapter 5 – Surface Treatment

5.1 General Review

As mentioned in the previous chapters, titanium alloys have a high specific strength. This desirable property has made them the metal of choice in the aerospace industry. They are often used as the main structural materials for aircrafts and spacecrafts. In recent years, the application of titanium alloys in the petroleum and chemical industries has been promoted due to the advantages of their corrosion resistance, thermal stability and superior performance at low temperatures. In these fields, the components, such as blades, valves, pumps and pipelines, that may be subject to corrosive environments or severe temperature conditions, are made with titanium alloys. The drawbacks of titanium alloys are the relatively low wear resistance and hardness, which limit their use and reduce the working life (Lin and Du 2014).

1) The hardness of titanium alloys is about HV 150-350 MPa. Such low hardness value in many cases does not meet the requirements of practical application.

2) Titanium alloys have a low anti-friction property, which can be attributed to two main factors, namely:

- low plastic shear resistance and work hardening rate; and

- weak protection of the surface oxide film.

The oxide film on titanium alloys is easy to peel off through the frictional contact, leading to a decrease in wear resistance. If a titanium alloy was put to use in a severely corrosive environment and if crevice corrosion was to occur, its corrosion resistance would be greatly reduced.

3) Due to low wear resistance, fretting wear would lead to a rapid decline of the fatigue strength of titanium alloys, which would make them difficult to use in sliding parts of mechanical products and automobile components.

The use of surface coatings is an efficient method for protection of titanium alloys against wear and corrosion (Bloyce et al 1998, Bansal et al 2011, Wang et al 2003). These coatings can be prepared through the following surface treatment technologies:

1) Traditional surface technologies, incorporating electroplating and thermal diffusion; and

2) Modern surface technologies, utilizing plasma, ion beams and electron beams.

Each surface treatment technology has its own inherent characteristics, and should be selected based on the operating requirements and conditions of titanium alloys.

In this section, a review of the application of diffent surface treatment methods in titanium alloys, encompassing thermal diffusion, anodic oxidation, micro-arc oxidation, laser cladding, ion implantation, electroplating and electroless plating, is provided.

5.1.1 Diffusion Coatings

Thermal diffusion is a process in which a metal or non-metal material is penetrated into a base metal at elevated temperatures in a controlled chamber (Corrosionpedia 2016, Ding and Wang 2006). The penetration layer is firmly bonded with the base metal through forming the alloy. An alloying layer with different composition and different thicknesses can be obtained by choosing different material as the source and suitable technological conditions. Nitride (TiN) and carbide (TiC) films on the surface of titanium alloys are normally prepared using nitriding, carburizing or carbonnitriding techniques to improve their corrosion resistance in highly corrosive environments, resistance to oxidation in high temperature conditions, hardness and wear resistance.

5.1.1.1 Nitriding

During nitriding, nitrogen is diffused into the titanium surface to create a case-hardening layer (TiN) (Zhuecheva et al 2005). TiN is hard and conductive. The heat of formation of TiN exceeds all titanium oxides. Therefore, completely removing oxygen is necessary to perform the nitriding treatment. The reaction between titanium and nitrogen follows the parabolic law, demonstrating a decrease in the nitriding rate with an increase of time. It is not possible to produce a thick nitrided layer since the diffusion rate of nitrogen in TiN outer layer of is slower than that in titanium solid solution inner layer. The content of oxygen and water must be strictly controlled in nitrogen or ammonia, to avoid hydrogen absorption and oxygen interference in the formation of nitrided layer.

Temperature increase can facilitate the growth of nitrided layer. At a temperature of 1123 K (850°C), the nitrided layer with a thickness of 25-1000 μm and a hardness of HV 3000 MPa is produced after a period of 16-

24 h. However, a higher temperature easily leads to the growth of grains in the titanium substrate, thereby affecting the substrate properties. For Ti-5Al-2.5Sn, the nitrided layer with a hardness ≥HV 5000 MPa and a thickness of about 50 μm can only be obtained after 16 h at 1123 K. Under the same conditions, the use of ammonia for nitriding can achieve a higher surface hardness and a larger hardening depth. But hydrogen absorbed in this process must be eliminated by vacuum annealing.

Different compositions may be formed during nitriding. If the oxygen content is not high, a titanium nitride outer layer is generated, which is golden yellow and has a hardness of HV 14-17 GPa, but is very thin (less than 5 μm). At lower nitriding temperatures or after subsequent annealing at high temperatures, this thin layer no longer increases or disappears due to the complete dissolution of nitrogen into the titanium solid solution on the surface of titanium substrate. A zone of obvious hardness drop in the case-hardening layer can be observed (Figure 5-1). In terms of α-β titanium alloys, the differences in the solubility and diffusion rate of nitrogen in the α and β titanium solid solution cause uneven thickness of the hardening layer.

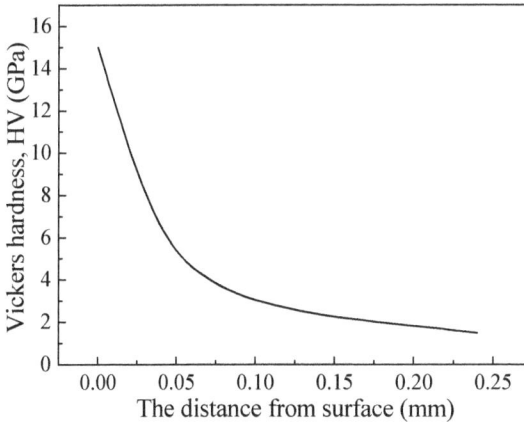

Figure 5-1 The hardness of nitriding layer vs. the distance from surface (After Ding and Wang 2006)

It should be noted that a thin nitrided case may be removed at any time, or in the presence of compressive stress or impact stress under severe and harsh conditions. Accordingly, it is essential to have a sufficiently thick nitrided layer to improve the wear resistance of titanium substrate. The corrosion resistance of titanium alloys in a reducing medium can also be improved by nitriding.

5.1.1.2 Carburizing

A combination of titanium and carbon can generate a stable carbide with high hardness. The growth of carburized zone between β-Ti and carbon is dependant upon the diffusion rate of titanium in the carburized zone. The diffusion coefficient D of titanium in carbide at 1561 K (1288°C) and 1761 K (1488°C) is 2.0×10^{-10} cm/s and 2.0×10^{-19} cm/s, respectively. The solubility of carbon in α-Ti is small, a total of 0.3% at 1123K (850°C), and declines to about 0.1% at 873 K (600°C). Because of the small solubility of carbon in titanium, surface hardening is achieved mainly through the titanium carbide layer and the subjacent deposition layer. It is different from the carburizing of steel in that carburizing of titanium alloys must be carried out under deoxidation, otherwise the carburized layer is prone to spalling. The carburized depth should be properly controlled, as an increase of depth causes the titanium carbide to become more brittle, and easier to peel off. The surface hardness of Ti-6Al-4V can increase by nearly 3 times after carburizing treatment (Figure 5-2).

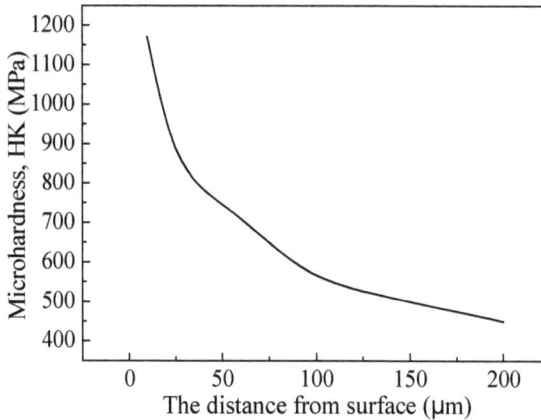

Figure 5-2 The Knoop hardness of carburized layer on Ti-6Al-4V (After Zhao 2003)

5.1.1.3 Boronizing

The titanium boride layer is very hard and has extraordinary wear resistance. The two methods, solid boronizing and laser boronizing, are mainly employed in preparing the titanium boride layer on the surface of titanium alloys (Sarma et al 2012). For solid boronizing, it is completed for 5-20 h at 1000-1050°C. In a short time at low temperatures, the boronized depth is only 0.8 μm. But after a long time at elevated temperatures, it can attain 15 μm depth. As Al is insoluble in boride, it is hardly detected in the boronized layer. Owing to the small atomic radius of V, there may be a small amount of V solid-dissolved in the interstitial positions of titanium boride crystals. The Al and V contents increase where the boronized depth is in excess of 10 μm, and is higher than that of Ti-6Al-4V substrate. When the boronized depth is close to the substrate, the Al and V contents drop to the value equal to the substrate. This may be explained by a "squeezing effect", which is induced by the process when the B atom is bound to Ti to form boride. Finally, the Al and V enrichment region is formed between boride and substrate. The B in the boronized layer possesses a stable content. It has a smaller amount between the boronized layer and the substrate, and is hardly detected in the substrate.

The microhardness of the boronized zone from the surface to the substrate presents a gradient distribution (Zhao 2003). The outer layer has the highest hardness, about HV 22GPa, due to the formation of a hard phase TiB_2. The Vickers hardness from TiB_2 to TiB in the boronized zone ranges from 22 GPa to 11 GPa. The B atom reacts with Ti to form intermetallic compounds, leading to a high hardness and a good wear resistance in the boronized layer. The hardness value of the transition zone from the titanium boride layer to the substrate drops remarkably, but it is greater than that of the substrate due to the enrichment of Al, V and other alloying elements (the Vickers hardness of Ti-6Al-4V is 3.3-3.5 GPa under the same load). The hardness of Ti-6Al-4V coated with the boronized layer is increased by 3-6 times. The boronized layer adheres well to the substrate, and does not erode or abrade easily. Thus, boronizing is a practical and effective way to improve the wear resistance of titanium alloys.

5.1.1.4 Molybdenumizing

Molybdenumizing is implemented on a titanium alloy via an ion diffusion technology to improve its propterty. The composition of the molybdenumized layer is MoTi alloy. This layer can significantly raise the hardness of the titanium substrate (Figure 5-3). As can be seen from Figure 5-3a, after molybdenumizing at 1020°C, the MoTi alloying layer shows a hardening effect. The hardness of the alloying layer is about twice as high as the substrate. It is important to note that the dry friction coefficient of the molybdenumized layer is smaller than that of the substrate, therefore titanium alloys coated with the surface alloying layer show good wear resistance. Although the coefficient of friction for coated Ti-6Al-4V has some fluctuations within the friction distance in the range of 0-150 m, it is still approximately 50% lower than that of the uncoated Ti-6Al-4V (Figure 5-3b).

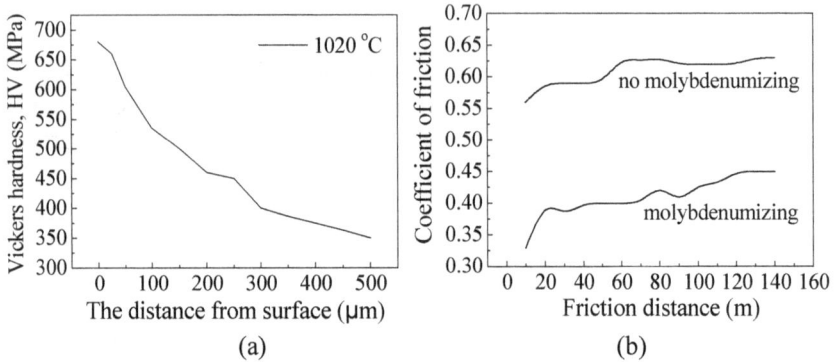

(a)   (b)

Figure 5-3 Microhardness (a) and coefficient of friction (b) before and after molybenumizing on the surface of Ti-6Al-4V (After Zhao 2003)

5.1.1.5 Oxidizing

In addition, in order to meet the requirements of environmental protection, surface oxidizing treatment technology has been paid more and more attention in recent years due to merits such as short treatment time, pollution-free and energy savings.

Oxygen, which is an interstitial element in titanium and also a stabilizing element of α phase, can enhance the allotropic transformation temperature of titanium and expand the α phase region. The maximum solubility of oxygen in α-Ti is up to 14.5wt%, while it is only 1.8% in β-Ti at 1740°C. In Ti-O solid solution, oxygen is present in the form of oxide in the crystal lattice of titanium, which enables the phase transition temperature to be increased significantly. Solid-dissolved oxygen in titanium can evidently improve the phase transition temperature from α-phase to β-phase and stabilize α-phase, while also increasing the microhardness due to an increase of the c/a value. After oxidizing, aside from a thin oxide film formed on the surface, supersaturated solid solution with a high hardness above 500 HV in titanium is the main reason for the microhardness improvement (Zhao 2003). Accordingly, this technology will have wide application prospects. At present, the mechanism of surface oxidizing treatment of titanium alloys has had a relatively short research time and there is still much to learn. Its application to an actual workpiece has just started, so this technology needs further study.

5.1.2 Anodic Oxidation

Anodic oxidation (anodizing) is an electrochemical process in which an oxide film on a metallic substrate is produced. It involves the application of an electrical bias and current while the substrate acts as an anode and is immersed in an electrolyte bath, as illustrated in Figure 5-4.

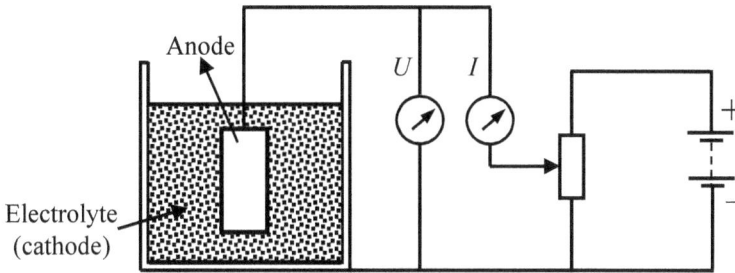

Figure 5-4 Schematic diagram of anodic oxidation system

The oxide film, which is formed through high temperature exposure in air (higher than 650°C), is unstable and easily comes off from the titanium substrate. This results in an uneven rough surface. But in contrast, the oxide film generated by anodic oxidation is similar to that formed spontaneously upon exposure to the air. It is dense and cohesive, but much thicker than that formed naturally (Simka et al 2011).

There are some problems in the application of titanium alloy:

1) Galvanic corrosion occurs when a dissimilar material is coupled with a titanium alloy;

2) It is susceptible to tri-corrosion under medium and high loads.

The way to overcome these problems is to modify the surface properties of the titanium substrate. Among the surface treatment methods of titanium alloys, anodic oxidation is simple and the oxide film produced has good resistances to wear, galvanic corrosion, fretting corrosion and hydrogen embrittlement.

Anodic oxidation of titanium alloys falls into two categories:

– functional anodic oxidation for improving corrosion resistance and wear resistance of the substrate;

– decorative anodic oxidation for altering the appearance of the substrate with distinctive tone.

In order to obtain a satisfactory coating quality, the following factors should be considered:

a) Electrolyte bath

Three kinds of baths, aqueous solution, organic solvent and water/organic solvent, can be used as anodic oxidation electrolytes of titanium alloys. A thin titanium oxide film of high dielectric, which has large resistance, high electrostatic capacitance, high leakage capacitance and low leakage current, is created in non-aqueous solutions and molten salt electrolytes. In aqueous solutions, the electrochemical sediments generated by electrolysis easily gather to form larger particles, leading to the growth of a thick porous oxide film. This film is a type of interference chromogenic coating, which plays a role in decoration and corrosion protection. Sulphuric and phosphoric acids are the most commonly used electrolytes to anodize titanium alloys.

b) Power supply waveform

Different modes of power supply can also affect the growth speed, the thickness and quality of the film. Four power supplies with different waveforms, full wave rectifiers, pulse direct-current, half wave rectifiers and smooth direct-current, were used to investigate the effect of power supply waveform on the performance of the anodic oxidation film of Ti-6Al-6V-2Sn-0.5Cu-0.5Fe in electrolytes containing H_3PO_4, $H_2C_2O_4$, and $KMnO_4$. It was found that the colored film obtained by pulse direct-current power supply had the best salt fog corrosion resistance, acid corrosion resistance, adhesion and wear resistance. When a half wave rectifier was applied, the oxide film showed the worst performance. The oxide film on the titanium surface prepared by a pulse direct-current power supply was dense, thick, and it had high hardness, good insulation and no chalking.

c) Process parameters (bath temperature, current density and electrical bias)

Electrolyte concentration, bath temperature and current density have a small impact on the growth of a titanium oxide film (Diamanti and Pedeferri 2007). The applied electrical bias is the major influencing factor. The chromogenic tone varies with the thickness of the film, which is proportional to the bath voltage. So, the relationship between the bath voltage and the interference colors can be established. Accordingly, the thickness of the oxide film can be adjusted through the bath voltage, and then the chromogenic tone can be precisely controlled, which allows the titanium surface to appear the corresponding color (yellow, green, pink, etc).

The current anodic oxidation processes of titanium alloys are presented in Table 5-1.

Table 5-1 The current anodic oxidation processes of titanium alloys (After Zhao 2003)

Anodic oxidation process		Characteristics of oxide film
Pulse anodic oxidation		
H_2SO_4	370-390 g/L	2 μm; gray color; good wear resistance and corrosion resistance; anti-viscosity.
H_3PO_4	16-32 g/L	
T	0-10°C	
S_c/S_A	2:1	
p_d	0.15-0.30 s	
p_t	40-120 /min	
i	3-7 A/dm^2	
Oxalic acid anodizing		
$C_2H_2O_4$	55-60 g/L	0.2-0.3 μm; light gray color; a certain degree of corrosion resistance and wear resistance; better wear resistance if anodizing under 8°C.
pH	0.5-1.0	
T	18-25°C	
S_c/S_A	1:10	
Thick film anodic oxidation		
H_2SO_4	350-400 g/L	20-40 μm; color change from gray to dark gray with an increase in thickness; high hardness; good absorbability due to porous structure; better wear resistance by adding colloid graphite and dry film lubricant into bath.
HCl	60-65 g/L	
T	40-50°C	
i	2-4 A/dm^2	
Colored anodic oxidation		
Formula 1:		**Formula 1**: If the voltage is gradually raised from 5 V to 50 V within 15 min, the color of oxide film is light brown, and then blue purple and finally golden yellow; if within 1-2 min, voltage rises to 50 V and is maintained 15 min, the color is changed from light blue to golden yellow.
CrO_3	120-150 g/L	
H_3BO_3	3-5 g/L	
T	18-25°C	
Formula 2:		
H_3PO_4	50-200 g/L	**Formula 2**: If within 20 min the voltage gradually rises from 5 V to 9 V, different color is obtained in different voltage range.
Organic acid	20-100 g/L	
T	18-25°C	

5.1.3 Micro-arc Oxidation

Micro-arc oxidation (MAO) is a new green technology for the growth of a ceramic coating on the surface of a metal. It is developed on the basis of anodic oxidation. The MAO system is illustrated in Figure 5-5. Ti parts are electrically connected, acting as one of the electrodes in the electrochemical cell, with the other "counter-electrode" typically being made from an inert material such as stainless steel (SS), and often consisting of the wall of the bath itself.

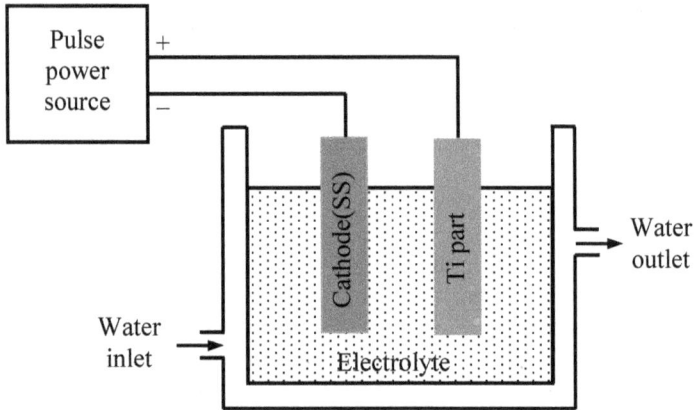

Figure 5-5 Schematic diagram of MAO system

However, there are many differences in film formation mechanisms and film performances between MAO and anodic oxidation. MAO is done by creating micro-discharges on the surface of materials immersed in an electrolyte at a higher applied potential than anodic oxidation, at least 200 V. A physically protective oxide film of high porosity is produced by subsurface oxidation, which is much thicker than anodic oxidation. It strongly adheres to the substrate and exhibits high hardness as well as good uniformity. In addition to high film quality, MAO offers a wide range of color pigmentation and environmental friendliness. It can help titanium alloys acquire a better corrosion resistance, pyro-oxidation resistance, wear resistance and insulating property.

The composition, structure, morphology and physical and chemical properties of the coating are highly dependent on electrolyte composition, electrical parameters (current density, pulse frequency, current density ratio of cathode to anode), as well as the substrate used. The performance of the oxide film prepared with AC power supply is better than prepared with DC power supply, so the AC mode is an important trend of development in MAO technology.

A MAO coating on a titanium alloy is basically composed of rutile TiO_2 and anatase TiO_2. Both rutile and anatase have a tetragonal structure. Rutile has a melting point of 1870°C, which is very stable under various temperatures and pressures. Anatase is a metastable phase, and can be transformed into rutile when heated. The electrolyte system of MAO mostly applied in a titanium alloy is alkaline, mainly including silicate, aluminate, phosphate (Wang et al 2006). The phase composition and content in the coating obtained using different electrolyte systems are slightly different. The coating formed in silicate electrolyte contains a large amount of SiO_2 amorphous phase and a small quantity of anatase TiO_2 phase. The rutile content in the inner layer is higher than in the outer layer. If aluminate electrolyte bath is used, the major composition in the outer layer of the coating is crystalline $TiAl_2O_5$ and a small amount of rutile TiO_2. The inner layer has much less $TiAl_2O_5$ and much more rutile. Phosphate electrolyte gives a higher content of anatase TiO_2 in the coating due to the addition of potassium dichromate, resulting in an improvement of pitting resistance.

The coating has a porous structure, and the diameter of the holes which are similar to craters are usually around 1-2 μm. These holes are the discharge channels of the plasma. The transition layer, compact layer and loose layer constitute the MAO film on the surface of titanium alloys. The transition layer adheres well to the substrate by concave-convex lapping. The compact layer without cracks is made up of fine equiaxed grains. It adheres tightly to the transition layer. Small cracks exist in the loose layer, but are not present throughout the entire layer. MAO film forms a continuous barrier on the titanium substrate.

On the whole, the noticeable characteristics of MAO oxide film include (IBC Coatings Technologies INC. 2016, Fei et al 2009):

- controllable thickness;

  The thickness of MAO coating is as low as 1 μm and as high as 250 μm.

- controllable color;

- significantly less wear compared to anodizing;

- low coefficient of friction and high hardness;

  The microhardness of MAO coating is HV 1000-2000 MPa, the highest hardness is HV 3000 MPa. This is comparable to hard alloys and is much higher than that of heat-treated high-carbon steel, high-alloy steel and high speed tool steel.

- significantly better corrosion resistance compared to anodizing;

- good electrical insulating property and pyro-oxidation resistance;

  The insulation resistance is up to 100 MΩ.

- strong binding force with the substrate;

- compactness and uniformity.

5.1.4 Laser Cladding

Laser cladding is a method of adding one material to the surface of another through a rapid melting and solidification process achieved by a laser with high power density. Based on the supply mode of the chosen cladding material, two classes are recognized: preset laser cladding and synchronized laser cladding. For the former, the chosen material is placed where it is desired on the substrate surface prior to cladding and then a focused laser beam is scanned across the target surface and melts the applied material to form a melt pool, leaving behind a deposited coating with good adhesion. The latter is to feed the cladding material to the laser beam directly. The feeding and cladding proceed concurrently (Figure 5-6). The cladding material is added in the form of powder, wire or plate.

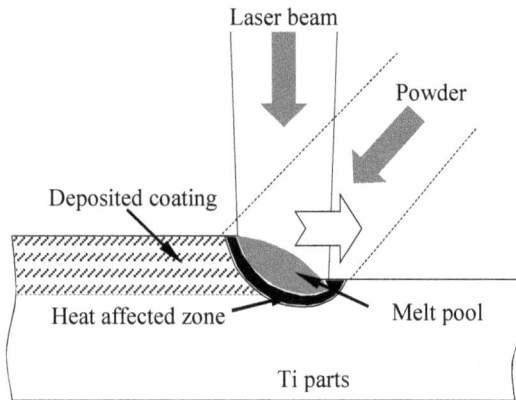

Figure 5-6 A schematic of laser cladding

This technique has a very wide choice of cladding materials and is suitable for a variety of substrate materials (metal, ceramic, even polymer). It is often used to improve mechanical properties or increase corrosion resistance, repair worn-out parts, and prepare metal substrate composites (Laser Cladding Technology Ltd 2016). High heating and cooling rate results in:

- deposits having fine microstructure and little or no porosity;
- narrow heat affected zone;
- limited distortion of the substrate and no need for additional corrective machining.

5.1.4.1 Cladding Materials

Self-fusing alloy materials, composite materials and ceramic materials are most widely used as the main laser cladding materials on the surface of titanium alloys (Niu and Sun 2006).

1) Self-fusing alloy materials

There are three series: iron-based alloy, nickel-based alloy and cobalt-based alloy. They all contain boron and silicon elements which have the effect of strong deoxidization and self-fusing. In the course of laser cladding, boron and silicon are oxidized to produce an oxide, and then form a thin film on the surface of the cladding layer. This film can prevent excessive oxidation of alloy elements, and combine with the oxide of these elements to generate boron silicate slag, thereby reducing inclusions and oxygen content. Moreover, boron and silicon can also decrease the melting point of the alloy, improve the wettability of the fuse-element on the substrate, and have a favourable effect on the fluidity and surface tension of the alloy. This is helpful to form the coating with low oxide content, low porosity and good adhesion.

The comparison of self-fusing alloy powder system series is listed in Table 5-2.

Table 5-2 Comparison of self-fusing alloy powder system

Series	Self-fusing performance	Merits	Demerits
Fe-based alloy	poor	low cost	poor oxidation resistance
Ni-based alloy	relatively good	excellent resistance to high temperature, good resistance to heat shock, good corrosion and wear resistance	high cost
Co-based alloy	good	good ductility and impact resistance, good heat and oxidation resistance, a high degree of corrosion resistance	relatively poor resistance to high temperature

2) Composite and ceramic materials

When a titanium alloy surface is subjected to sliding, severe impact and abrasive wear, using only self-fusing alloy materials may be insufficient. Under these circumstances, ceramic particles of carbide, boride and oxide with high melting points, are added into the self-fusing alloy materials to produce metallic ceramic composite coatings.

The most studied and applied materials are carbides (WC, TiC, SiC, etc.) and oxides (ZrO_2 and Al_2O_3, etc.). The duration of the melt pool in laser cladding is short at elevated temperatures, so the ceramic particles are unable to completely melt. The deposited coating consists of face-centered cubic γ phase (Fe, Ni, Co), unmelted ceramic particles and precipitated phase (such as MC). The multiple strengthening mechanisms, refined crystalline, hard particle and solution and dislocation accumulation, contribute to the increase in the hardness and wear resistance of the deposited coating.

The properties of the deposited coating are determined by the microstructure and phase composition, which are strongly dependent on the processing technology. Hence in the selection of a cladding material, besides the basic requirement of laser cladding for the cladding material (e.g., the deposited coating obtained can possess good wear resistance, corrosion resistance, high temperature resistance and oxidation resistance), good technological performance of the cladding material must be considered, including the following aspects:

a) thermal expansion coefficient match between the cladding material and the titanium substrate.

One of the most important causes of cracks in coating is the difference in the thermal expansion coefficient between the cladding alloy and the titanium substrate. Generally, the thermal expansion coefficient of the coating should be as close as possible to the substrate in order to prevent cracking and spalling of the coating.

b) melting point match between the cladding material and the titanium substrate.

Another important thermal physical property of cladding material is its melting point. If the melting point difference between the cladding material and the substrate is too large, a good metallurgical bond can not be formed.

c) wettability of the cladding material on the titanium substrate.

Wettability is relevant to surface tension. The small surface tension enables the coating melt to uniformly spread on the titanium surface, that is, the melt has better wettability. The good wettability can result in the formation of a homogeneous coating. The basic principles to enhance the wettability of the coating:

- reducing the surface tension of the coating melt;
- reducing the surface tension of the substrate;
- reducing the solid/liquid interfacial energy between the coating and the substrate. Specific measures include increasing the cladding temperature and adding active elements, etc.

5.1.4.2 Cladding Technology

The quality of laser cladding is closely related to three parameters: laser power P, laser beam diameter D and scanning speed V (Niu and Sun 2006).

1) Laser power

When the other parameters remain unchanged, a larger laser power can melt more titanium substrate materials. However, there is a higher possibility of producing pores. This can be mitigated by increasing the laser power, which causes the depth of the cladding layer to increase, then the metal liquid flows to the pores, and as a result, the pores are gradually reduced and even eliminated. When the cladding layer depth reaches the limit value, a further increase in power enlarges the plasma, and then makes the temperature on the substrate surface go up. This leads to the aggravation of deformation and cracking. On the other hand, if the laser power is too low, only the surface coating is melted, but the substrate is not melted. We can observe the phenomena of local balling and cavities in the cladding layer. In this scenario, the coating with the desirable properties is unobtainable.

2) Laser beam diameter

The laser beam is generally circular. The width of the deposited coating mainly depends on the diameter of the spot. The cladding layer is widened with an increase in spot diameter. At the same time the surface energy distribution of the cladding layer varies with spot size. These factors cause the difference in morphologies and mechanical properties of the deposited coating. Normally, good quality is obtained when the spot size is small. The quality decreases with an increase in spot size. But if the spot diameter is too small, it is not conducive to gain the cladding layer in a large area.

3) Scanning speed

The scanning speed has a similar effect as the laser power. When it is too high, the powder can not be completely melted; if it is too low, the existence period of the molten pool is too long and the powder is overburned, resulting in loss of alloying elements. Simultaneously, a large heat input into the titanium substrate increases its distortion.

Most of the research found that the laser cladding variables are not independent. They are interrelated in determining the macroscopic and microscopic quality of the deposited coating.

5.1.4.3 Categories of Cladding Coating

In regard to the types of laser cladding coating on titanium alloys, there is wear resistance coating, corrosion resistance coating, anti-oxidation coating, thermal barrier coating and bio-ceramic coating, and so on (Tian et al 2005).

Wear-resistant coating is the first and most extensively studied in laser cladding. There is some literature reporting research on the microstructure and properties of a coating prepared on the surface of a titanium alloy using cladding materials, NiCrBSi and NiCrBSi+TiC powder. The coating had a hardness of HV 900-1200 MPa and a better wear resistance than the substrate (Guo et al 2010).

The structural alloys exposed to an oxidizing atmosphere often need an anti-oxidation coating to prolong their service life. Ti4Si6Ni80 alloy powder was pre-coated with high strength titanium alloy Ti-6.5Al-1.5Zr-3.5Mo-0.3Si at elevated temperature, and then was melted by laser beam. The obtained newly intermetallic compound composite coating was made up of intermetallic compound Ti_5Si_3, NiTi with a small amount of Ni and solid solution γ. This kind of coating presented a good oxidation resistance at elevated temperatures, which was 1.8 times higher than that of the substrate.

To sum up, laser cladding can improve the surface properties of titanium alloys and acquire coatings with a high hardness, good wear resistance, good high-temperature oxidation resistance and low friction coefficient.

5.1.5 Ion Implantation

Ion implantation is a physical process in which positive ions are accelerated in a high-voltage electrical field and impacted into a solid material surface at a high speed. The typical implanted depth in the target is in the range of tens of angstrom to thousands of angstrom, which is governed by implanted ion mass, implanted ion energy and target material type. The implanted ions are produced by ionization of gas or metallic vapour, which can cause many chemical and physical changes in the target by transferring their energy and momentum to the electrons and atomic nuclei of the target material.

To increase wear resistance and fatigue property of a titanium alloy, commonly C ions or N ions are implanted to the surface to form TiC or TiN in the atomic scale depth of the outermost layer of its surface. Many researchers reported that N ion implantation modified the surface composition and microstructure, which had a good effect on hardness and tribological properties of Ti-6Al-4V (Budzynski et al 2006). If Yt ions are pre-implanted into Ti-6Al-4V before N ions are implanted through plasma immersion, it is beneficial to increase the implantation depth of N ions and significantly improve the wear resistance of Ti-6Al-4V. TiC is also a super hard phase, so carbon ion implantation can also strengthen titanium alloys. In some cases, a single carbon deposition layer with similar structure as diamond, that is, "diamond-like carbon" (DLC), can form on the surface. This modified layer has a lower friction coefficient and a better wear resistance than a nitrogen implantation layer. If N and C ions are implanted together, the wear resistance of the modified materials can increase by nearly 3 times. The friction and wear resistance and cyclic fatigue resistance can be improved when Ba ions are implanted to titanium alloys.

Ion implantation has the following advantages over other surface treatment methods:

1) The film adheres well to the substrate and presents strong resistance to mechanical and chemical actions;

2) There is no need to raise the substrate temperature during the implantation process, which can maintain the geometric precision of the workpiece;

3) The technology has good repeatability.

5.1.6 Electroplating

Due to its fast deposition rate, relatively low process temperature, and reusability of the plating bath, electroplating is a common method of depositing metals or alloys on titanium alloys.

In this method, a titanium alloy acts as a cathode and another metal is employed as an anode in a conductive aqueous solution containing the anode metal salts (Figure 5-7). The power supply provides a direct current to the soluble anode, causing it to dissolve to produce metal ions for the aqueous solution and electrons for the power supply. Simultaneously, the metal ions in the solution are attracted to the cathode and gain electrons from the power supply to form a deposit layer on the surface of the cathode. However, if an acid solution is used, hydrogen ions may also gain electrons from the cathode to form hydrogen gas bubbles. This is not good for the plating process and the titanium substrate. It lowers plating efficiency and may cause poor deposit properties. Moreover, the substrate absorbs hydrogen and may suffer from hydrogen embrittlement.

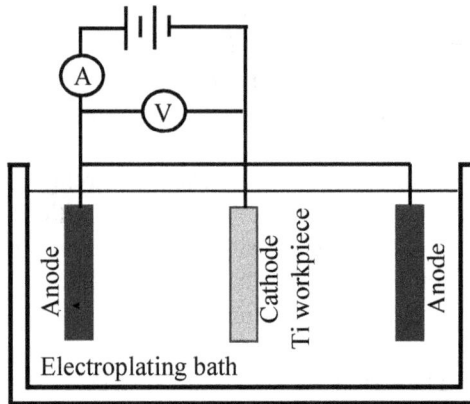

Figure 5-7 Electroplating setup

Some electrodeposited coatings applied in a titanium substrate are as follows (Kim et al 2006):

1) Nickel coating can enhance the corrosion resistance of the titanium substrate and make it less susceptible to wear.

2) Chromium coating is favorable for reducing hydrogen absorption of the titanium substrate, and significantly increasing the surface hardness and wear resistance.

3) Fe and Fe-Ti alloys are electrochemically deposited on the titanium substrate for achieving high hardness, high strength, good high-temperature properties and good fretting wear resistance.

4) Copper is plated on the titanium substrate for the purpose of improving the surface condition to prevent thread galling. This is vital for titanium parts connected by mechanical methods, such as compressor

blades. Copper can also be plated to improve electrical conductivity.

5) After the pretreatment, including sandblasting, acid pickling and copper pre-plating, silver coating with a good adhesion and a thickness greater than 25 µm is obtained on the titanium substrate to enhance electrical conductivity and solderability.

6) Gold can be plated on the titanium substrate in aerospace industry as a means of increasing surface reflectance.

5.1.7 Electroless Nickel Plating

The electroless nickel plating technique has a long development history. In 1944, German chemists, A. Brenner and G. Riddell, successfully obtained a nickel phosphorus (Ni-P) alloy coating in a stable bath, bringing electroless nickel plating into a new stage. Extensive research on this new method of plating nickel has been performed in many countries, and it wasn't applied in industry until the late 1950s. Currently electroless nickel plating is an established industrial process (Mudali and Raj 2009).

In the electroless nickel plating process, the chemical reduction of nickel ions occurs in an autocatalytic way without the application of an external electric current. The desirable properties of the coating can be obtained at both simple and complex shape parts through effective control of the solution. Electroless Ni-P coating offers distinct advantages, such as good corrosion and wear resistance, low friction coefficient, high hardness, good solderability and uniform thickness (Mahmoud 2009, Dadvand 2002). An electroless Ni-P bath generally contains a nickel salt as a source for nickel, a reducing agent to supply electrons for the reduction of nickel, complexing agents to control the free nickel available for the reaction, and buffering agents to resist the pH changes caused by hydrogen released during deposition. Heat increases the temperature of the bath and provides the driving force for the deposition process.

The deposition process of Ni-P coating can be described by the following steps:

(1) The reducing agent (sodium hypophosphite) is oxidized and atomic hydrogen is generated.

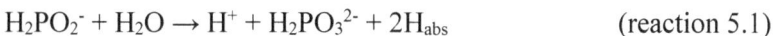

$$H_2PO_2^- + H_2O \rightarrow H^+ + H_2PO_3^{2-} + 2H_{abs} \qquad \text{(reaction 5.1)}$$

(2) Atomic hydrogen is absorbed and activated on the surface, and then the Ni ion is reduced to produce a Ni layer.

$$Ni^{2+} + 2H_{abs} \rightarrow Ni + 2H^+ \qquad \text{(reaction 5.2)}$$

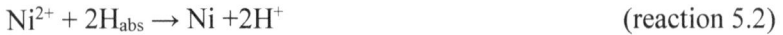

(3) Phosphorus and hydrogen gas are produced through further oxidation of sodium hypophosphite according to reaction 5-3 and 5-4:

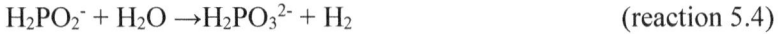

$$H_2PO_2^- + H_{abs} \rightarrow H_2O + OH^- + P \qquad \text{(reaction 5.3)}$$

$$H_2PO_2^- + H_2O \rightarrow H_2PO_3^{2-} + H_2 \qquad \text{(reaction 5.4)}$$

Ni-P alloy coating deposits by the simultaneous growth and integration of P and Ni present on the surface.

Currently, electroless plating on the surface of titanium alloys has been widely concerned. However, it has long been considered a difficult process. The main issue is that titanium is very reactive and when exposed to air, a thin oxide layer is formed on the underlying titanium substrate. This passive film can reduce the bond between the coating and the substrate, leading to poor adhesion, blistering and coating failure. Therefore, it is necessary to find a suitable activation method to improve adhesion between the Ni-P coating and the titanium substrate. Meanwhile, in spite of lots of work reported on electroless Ni-P process, no in-depth information presenting the process parameters applicable to titanium alloys and their influence on wear resistance of titanium alloys is currently available.

There is also a class of electroless nickel plating in which micro-sized powder particles are added into an electroless bath, known as electroless composite plating. The composited particles have two main types: hard particles (e.g. SiC, diamond) and lubricant particles (e.g. graphite, MoS_2, PTFE, BN, $(CF)_n$, mica). These composite coatings can provide unique wear resistance and offer increased serviceability, which are primarily used in aggressive abrasion and erosion wear applications.

MoS_2 is a solid lubricant with chemical stability. It is mainly applied in special environments where the use of a traditional liquid lubricants becomes ineffective or cannot be tolerated, such as in a hydraulic system and pump, a computer, nuclear industry, satellite components and aerospace parts. Therefore, electroless $Ni-P-MoS_2$ composite coating has begun to attract attention. Jin hui et al (2005) compared the wear resistance of Ni-P coating, Ni-W-P coating and $Ni-P-MoS_2$ composite coating, and found that the $Ni-P-MoS_2$ composite coating had the lowest wear loss and friction coefficient, the results of which suggested that MoS_2 can play a role in antifriction. At present, there are many papers discussing electroless composite coatings, but few studies mentioning MoS_2-dispersed electroless nickel composite plating on titanium alloys are reported. The important aspects in the studies of electroless $Ni-P-MoS_2$

composite coating on the surface of titanium alloys include: 1) increase of the MoS_2 content and homogeneity in the composite coating; 2) improvement of antiwear and antifriction performance of the composite coating; 3) co-deposition behavior of MoS_2 particles and Ni-P alloy.

5.2 Studies of Electroless Plating on Ti-6Al-4V

This section covers some of the research on electroless Ni-P coating (EN-P) and Ni-P-MoS_2 composite coating (EN-P-MoS_2) on Ti-6Al-4V (Lin and Zhao 2013, Lin et al 2013a, Lin et al 2013b, Wu 2014, Lin et al 2015a).

5.2.1 Experimental Details

The Ni-P coating and Ni-P-MoS_2 composite coating were prepared in an electroless plating device. The main components of the device were plating container, thermostatic water bath, thermometer, and stirrer (Figure 5-8). The specimen was hung vertically in the container which was placed in the thermostatic water bath and covered with a lid to prevent evaporation. The solution was stirred by a mechanical or magnetic stirrer, which enabled the temperature and concentration of bath constituents to remain uniform. Annealed Ti-6Al-4V with thickness of 2 mm was used as a substrate material for all experiments.

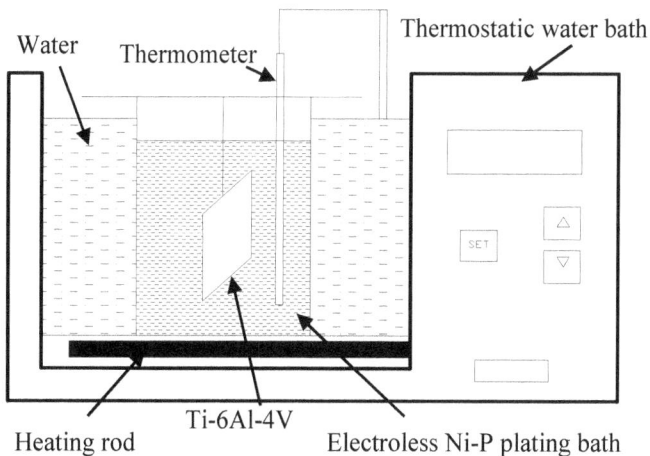

Figure 5-8 Experiment set-up of electroless plating

5.2.1.1 Electroless Ni-P Plating Procedure

The flow chart of the electroless EN-P coating and EN-P-MoS$_2$ coating on the surface of Ti-6Al-4V is depicted in Figure 5-9.

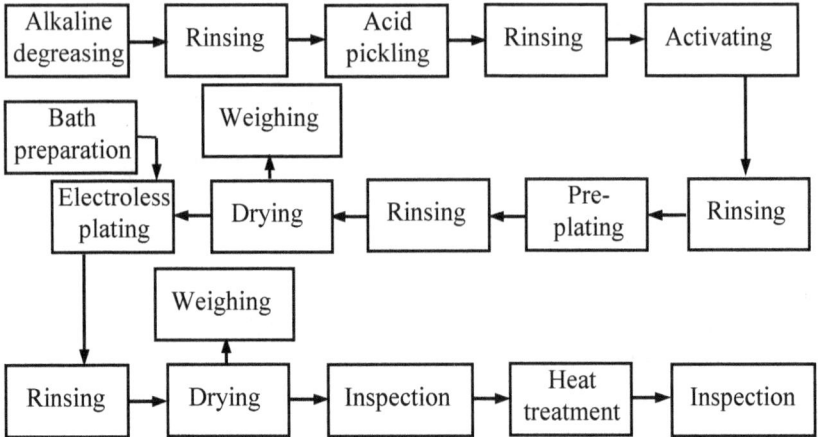

Figure 5-9 Flow chart of the electroless plating on the surface of titanium alloys

The pretreatment process, including alkaline degreasing, acid pickling, activating and preplating, is given in detail in the paper by Dr. Lin (2015a).

5.2.1.2 Performance Evaluation

After plating, an analysis of thickness, adhesion, morphology, composition, structure and wear resistance of the coating were carried out.

1) Deposition rate

The deposition rate is a speed at which a coating grows on the surface of a substrate. It can be defined by the increase in coating thickness per unit time.

$$v = \frac{\Delta m \times 10^4}{A \cdot \rho \cdot t} \tag{5.1}$$

where Δm is the mass gain (g), A is the specimen area (cm^2), ρ is the density of coating (g/cm^3), and t is the plating time (h).

2) MoS$_2$ content in the composite coating

MoS$_2$ content in the composite coating was obtained by the gravimetric method. The specimen was weighed, and then was dissolved in 1:1 nitric acid solution under stirring. It was removed from the bath and dried in air. The weight of the specimen was measured again. The weight loss, Δm, is the total weight of the composite coating. Following that, the precipitate in the solution was filtered, dried and weighed. The percentage of MoS_2 can be calculated from the ratio of the precipitate weight to the total weight of the composite coating.

3) Adhesive strength

The coating adhesive strength was examined by a scratch test. The critical load is the normal force applied on the scratch probe at the time when the coating is destructed, which is a good indication of the adhesion strength. Generally, a higher critical load indicates a higher adhesion strength. This can be seen from the curve of acoustic emission signal vs. load variation.

4) Morphology and microstructure

The morphologies and composition of the coating and the activation layer were observed using the same SEM and EDS equipment as mentioned in section 3.4.2.3. X-ray diffraction (XRD) was conducted to identify the phases of the coating employing a D_{MAX}-RB XRD system

5) Wear performance

The surface roughness of the coating was quantified by vertical deviations of a real surface. R_a was tested as the roughness parameter by JB-6C, and it is the arithmetic average of the roughness profile.

Hardness is a parameter characterizing the material local resistance against plastic deformation by the penetration of a hard body (indenter). It is of significance in evaluating wear and abrasion resistance for the Ni-P coating. A HVS-1000 Vicker microhardness tester was employed to measure the hardness.

A CFT-1 multifunctional material surface performance tester was adopted to investigate the coefficient of friction of the composite coating in rotational dry friction mode (Figure 5-10). During the test, a GCr15 steel ball with 5 mm diameter as a counter-face material rotated against a flat pre-weighed specimen with a rotating radius of 6 mm and rotating speed of 500 r/min for a period of 30 min.

5.2.2 Improvement of Coating Adhesion

The approaches to improve the adhesive strength of coating to the substrate includes activation before plating and heat-treatment after plating (Lin and Zhao 2013, Lin and Wu 2015a). In the activation methods, three chemical conversion coatings including double zincating film, tin film and phosphating film are considered.

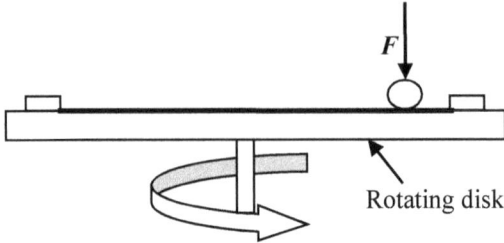

Figure 5-10 A graphical illustration of wear test

The preparation process of chemical conversion coatings is shown as below.

a)   double zincating film

HF (40 wt.%) 80 mL/L, $ZnSO_4 \cdot 7H_2O$ 15 g/L, $C_3H_6O_3$ 20 mL/L, room temperature.

First zincating 2 min → rinsing → film removal 30 s in 1:1 nitric acid → rinsing → second zincating 1 min

b)   tin film

NaOH 160 g/L, $Na_4P_2O_7$ 40 g/L, $NaAC \cdot 3H_2O$ 10 g/L, $NaSiO_3$ 40 g/L, pH=12, T=65°C, t=60 min.

c)   phosphating film

H_3PO_4 160 g/L, KH_2PO_4 4 g/L, $C_6H_{12}N_4$ 0.1 g/L, $NaNO_3$ 0.2 g/L, room temperature, t=75 s.

5.2.2.1 Chemical Conversion Coating

The test curves of adhesive strength between the EN-P coating and the titanium substrate using different chemical conversion treatments are shown in Figure 5-11. Under a 40 N load, the EN-P coating produced after the double zincating treatment is not damaged by diamond bit. This means that the coating can withstand the 40 N load or higher. The critical load is about 23 N for the EN-P coating on the titanium substrate activated by the tin film. During the test, parts of the coating peel off. The impact of the phosphating treatment is contrary to the expectation. It reduces the adhesive strength between the coating and the substrate, the critical load is only about 15 N and spalling of the coating is serious.

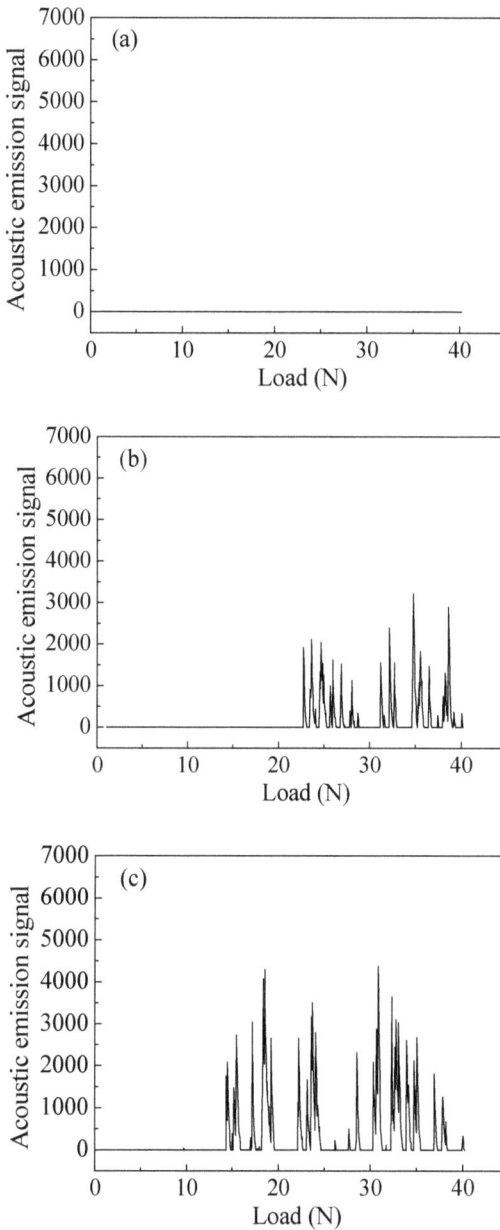

Figure 5-11 Adhesion test curves of EN-P coating on Ti-6Al-4V using different conversion treatment: (a) double zincating film; (b) tin film; (c) phosphating film

The images of cross section of the coating obtained using different chemical conversion treatments are shown in Figure 12.

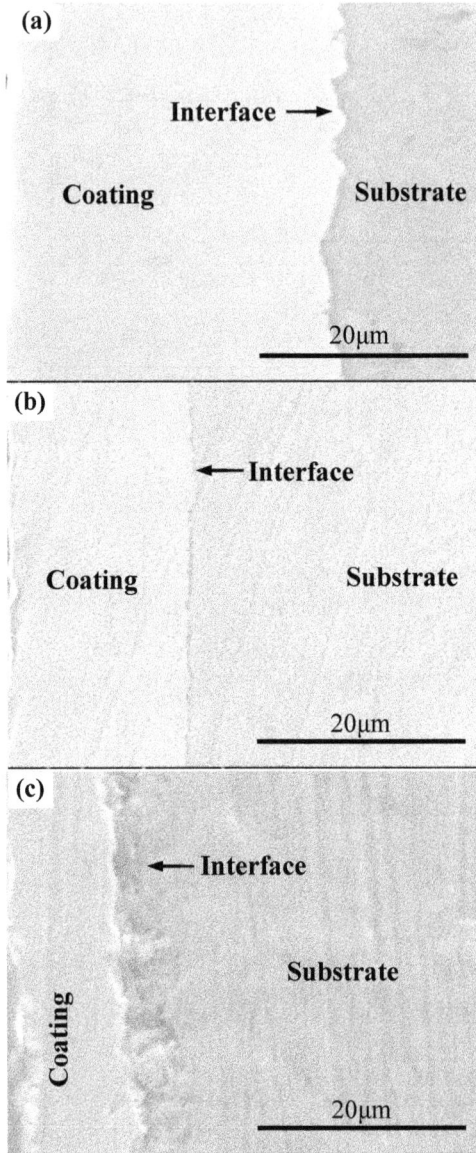

Figure 5-12 Cross section morphology of EN-P coating on Ti-6Al-4V using different conversion treatment: (a) double zincating film; (b) tin film; (c) phosphating film

The EN-P coating prepared after double zincating treatment appears to be well-bonded to the substrate (Figure 5-12a). No gap exists between the coating and the substrate. After the tin film is generated on the titanium substrate, there is a small gap between the EN-P coating and the substrate, which affects the adhesive strength (Figure 5-12b). After phosphating treatment, the bond between the EN-P coating and the substrate becomes poor, and there is a big gap between them (Figure 5-12c). By comparison, double zincating with better adhesive strength is the preferred activation method for electroless plating on Ti-6Al-4V.

5.2.2.2 Heat Treatment

Heat treatment after plating can not only reduce the stress in the deposited coating but also reduce the amount of hydrogen absorption accumulated. Furthermore, it can enhance the adhesive strength of the coating. The effect of the heat treatment temperature on adhesion of the coating is shown in Figure 5-13.

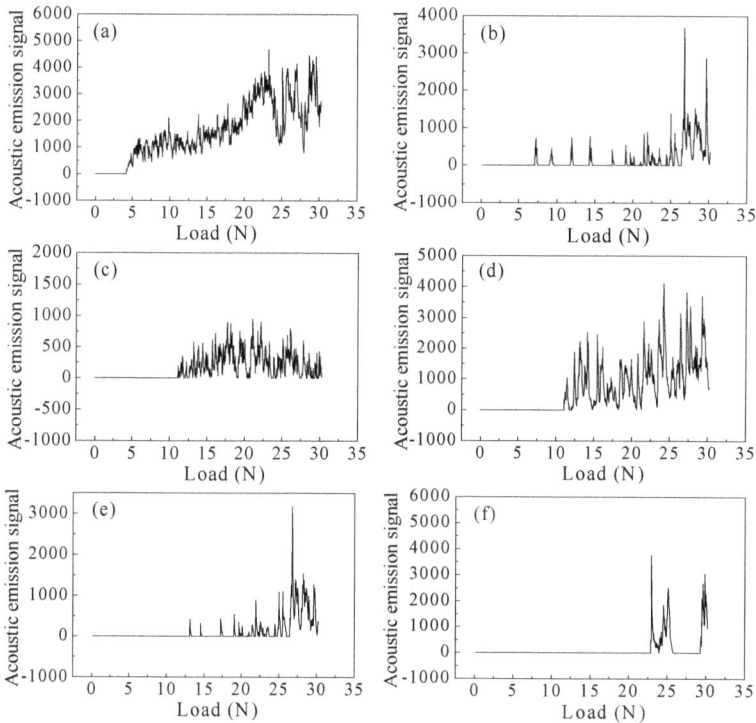

Figure 5-13 Adhesion test curves of EN-P coating on Ti-6Al-4V after heat-treatment at different temperatures: (a) 100°C; (b) 200°C; (c) 300°C; (d) 400°C; (e) 500°C; (f) 600°C

   Heat treatment at relatively low temperatures, 100°C and 200°C, only releases the internal stress and hydrogen in the coating. It has little effect on the adhesive strength. The critical load of the coating is about 25 N after heat treatment at 300°C. When the heat treatment temperature is raised to 600°C, the critical load reaches 35 N. High temperature treatments are propitious to mutual diffusion between the coating and the titanium substrate and then the formation of the diffusion layer with a certain thickness, which causes an increase in the adhesive strength of the coating. The cross-section morphology is illustrated in Figure 5-14. The interface between the coating and the titanium substrate indicates that the coating adheres tightly to the substrate and is evenly distributed by a thickness of about 20 μm.

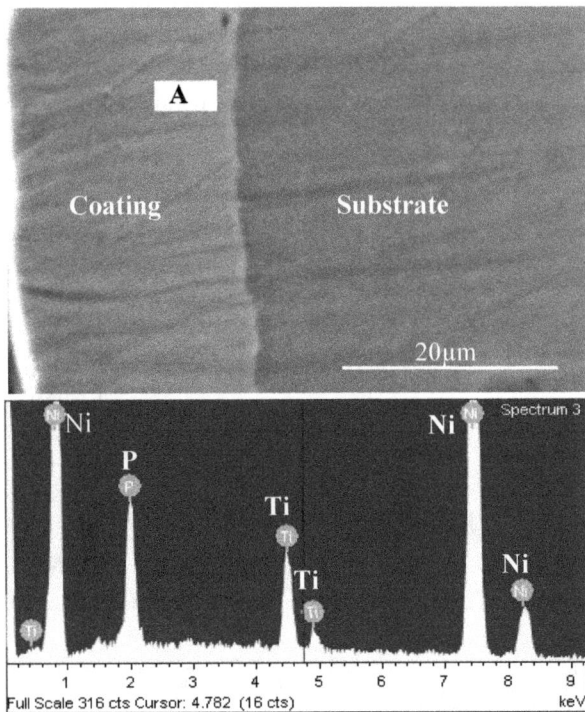

Figure 5-14 Cross section SEM image and EDS result of EN-P coating on the surface of Ti-6Al-4V after heat treatment at a temperature of 600°C

   The qualitative elemental analysis for spot A in the coating is also given in Figure 5-14. Ti is detected in the coating close to the interface between the coating and the substrate. This reveals that mutual diffusion of the atoms occurs between the Ni-P coating and the substrate after heat treatment, producing a diffusion layer. This is good for the improvement of cohesion.

5.2.3 Electroless Ni-P Plating

5.2.3.1 Preparation of Ni-P Coating

The major influential parameters on deposition rate and phosphorus content in the coating are discussed in this section for obtaining the optimal preparation process. This process needs to have the characteristics of high deposition rate (16-20 μm/min) and relatively low phosphorus content in the coating (4-10%) (Lin and Zhao 2013, Zhao 2012). The basic formula is nickel sulfate ($NiSO_4$) 0.1 mol/L, sodium hypophosphite (NaH_2PO_2) 0.2 mol/L, sodium citrate ($Na_3C_6H_5O_7$) 0.05 mol/L, lactic acid ($C_4H_6O_5$) 0.2 mol/L, malic acid ($C_4H_6O_5$) 0.05 mol/L, sodium acetate (NaAC) 0.1 mol/L, potassium iodate (KIO_3) 0.5 mg/L, thiourea (CH_4N_2S) 1.0 mg/L.

1) Molar ratio of $Ni^{2+}/H_2PO_2^-$

The effects of the molar ratio of $Ni^{2+}/H_2PO_2^-$ on deposition rate and phosphorus content are shown in Figure 5-15. Different molar ratio of $Ni^{2+}/H_2PO_2^-$ is obtained by varying the NaH_2PO_2 concentration and keeping the Ni^{2+} concentration constant (0.1 mol/L).

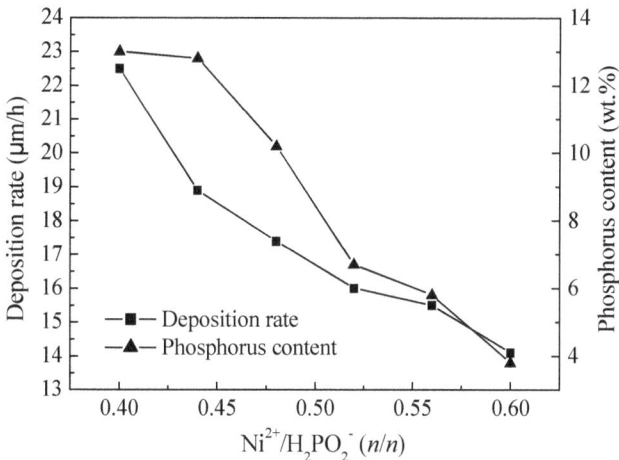

Figure 5-15 Deposition rate and phosphorus content in EN-P coating as a function of $Ni^{2+}/H_2PO_2^-$ ratio

With an increase in the molar ratio of $Ni^{2+}/H_2PO_2^-$, there is a decrease in deposition rate and phosphorus content. The reaction mechanism of electroless nickel plating shows that the deposition rate is reduced with a decrease in the concentration of the reducing agent i.e. NaH_2PO_2. Meanwhile based on the reaction of phosphorus production, a decrease in $H_2PO_2^-$ concentration can give rise to a drop in the phosphorus content. From the data in Figure 5-15, the optimum $Ni^{2+}/H_2PO_2^-$ ratio is within the range of 0.45 to 0.48 to ensure a relatively high deposition rate and a relatively low phosphorus content.

2) Complexing agent

A complexing agent has an important role in controlling the free nickel available for the reduction reaction, preventing the formation of phosphite precipitation, mitigating the decomposition of the plating bath and prolonging the service life of the bath. If the concentration of the complexing agent is too low, the bath is unstable and easy to decompose; if the concentration is too high, the deposition rate is low, and even the coating can not be deposited. The complexing agent also affects the phosphorus content in the coating.

Figure 5-16 depicts the effects of the concentration of the complexing agent on the deposition rate and phosphorus content. Sodium citrate and malic acid are 0.05 mol/L. Lactic acid varies from 0.05 mol/L to 0.3 mol/L.

An increase in the concentration of the complexing agent (lactic acid) results in an initial increase in the deposition rate followed by a decrease.

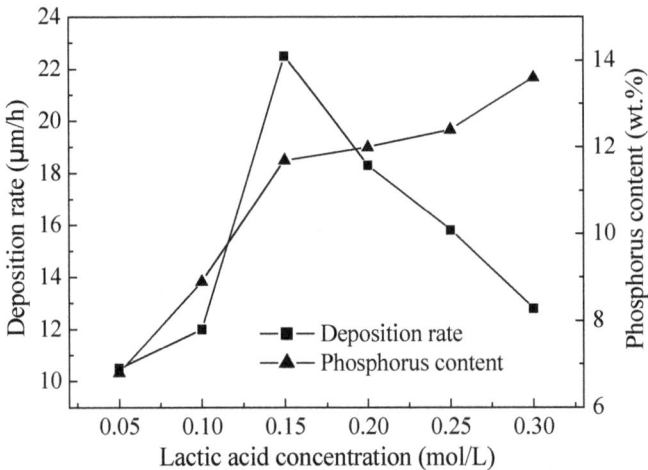

Figure 5-16 Deposition rate and phosphorus content in EN-P coating as a function of complexing agent concentration

The maximum deposition rate value occurs at a concentration of 0.15 mol/L. The phosphorus content in the coating continuously goes up as the concentration of the complexing agent increases. This is because the complexing agent not only regulates free Ni^{2+}, but also promotes the reduction of Ni^{2+}, but when the concentration of the complexing agent exceeds a certain range, it has a strong inhibition to the reduction reaction, leading to a rapid fall in the deposition rate. The concentration of lactic acid is controlled within a range of 0.15 mol/L to 0.2 mol/L.

3) pH

The pH value has a great influence on the electroless plating process, and it is an important factor that must be strictly controlled in the process parameters. If pH is too high, nickel phosphorous precipitation with a very small solubility is easily formed and the bath is easily decomposed; if pH is too low, the deposition rate is very slow. The corrosion and wear resistance of an electroless nickel coating are mainly related to its composition, in particular, the phosphorus content. The coating obtained at a lower pH value has a high phosphorus content and an amorphous structure, showing good corrosion resistance. At a higher pH value the deposited coating has a low phosphorus content and nanocrystalline structure, presenting good wear resistance. The effect of pH on the deposition rate and phosphorus content in the EN-P coating is depicted in Figure 5-17.

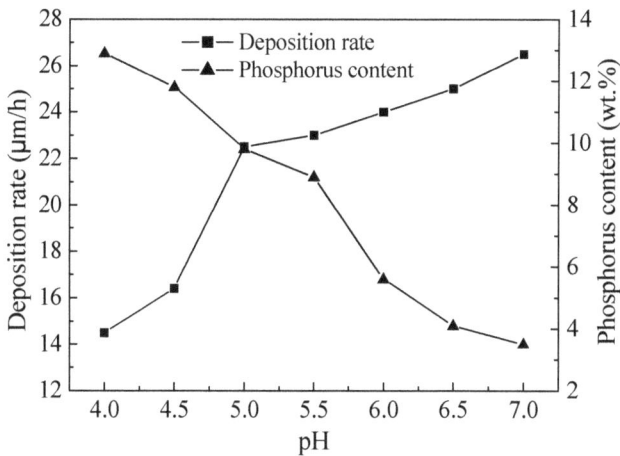

Figure 5-17 Deposition rate and phosphorus content in EN-P coating as a function of pH

As the pH increases within the range of 4 to 7, the deposition rate increases and the phosphorus content decreases. The selected pH value of the bath is between 5.0 and 6.5 at which the deposition rate and the stability of the plating bath can be guaranteed.

4) Temperature

Temperature plays an essential role in improving the deposition rate. High temperatures can promote the deposition process, but if the temperature is too high, the plating bath is unstable. When the temperature is below 40°C, the reaction becomes difficult.

The deposition rate vs. temperature curve is plotted in Figure 5-18. It increases with an increase in the temperature, which improves the diffusion ability of the reaction ions and the reaction activity. For acidic electroless plating baths, the selected range of temperature is from 85°C to 95°C. During electroless plating, it is maintained at ±2°C of this temperature range. This can ensure good coating quality, suitable deposition rate and bath stability.

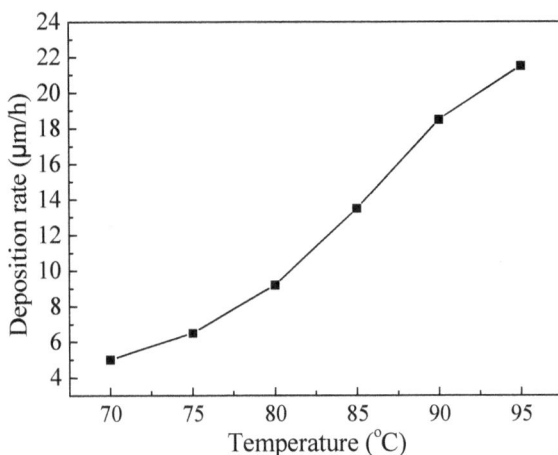

Figure 5-18 Deposition rate as a function of temperature

The relationship between the deposition rate and the reciprocal of temperature is shown in Figure 5-19. Using the Eq.(4.6), E_a can be obtained from the slope of the curve in Figure 5-19. For the bath formulation that is used in this work, the activation energy E_a is 7.01 Kcal/mol. The values reported in literature generally range from 11 to 23 Kcal/mol (Dadvand 2002). Compared with this, the activation energy in this work is lower, indicating that it is an easier deposition process to perform.

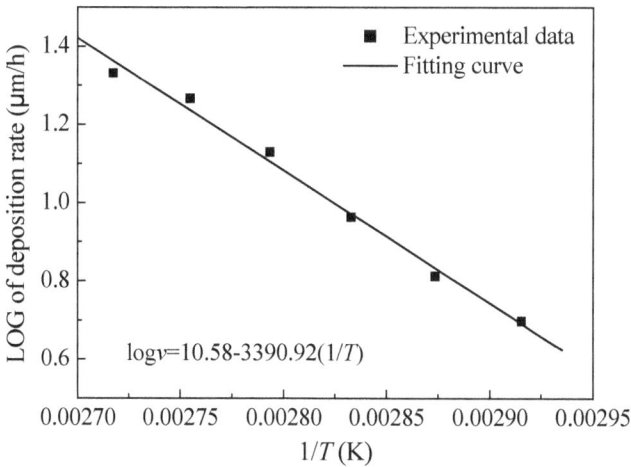

Figure 5-19 Deposition rate as a function of reciprocal of temperature

5.2.3.2 Microstructure of Ni-P Coating

A Ni-P coating with different phosphorus content can be obtained by changing the reducing agent NaH_2PO_2 concentration, the complexing agent concentration and the pH.

The SEM images and EDS results of the coatings with different phosphorus contents are demonstrated in Figure 5-20. The bath developed for generating the Ni-P coating on Ti-6Al-4V produces a uniform deposition. The surface morphology and microstructure of the coating is changed with an increase in phosphorus content. The coating with a low phosphorus content has a cellular structure which contains uniformly deposited nodules. These nodules are expected to provide a low friction surface by greatly reducing the contact surface. In the coating with a high phosphorus content, there are little or no nodules, also no grain boundaries. This enhances corrosion resistance.

The XRD spectra of the Ni-P coating containing 3.25%, 6.85% and 11.76% phosphorus are depicted in Figure 5-21. This can be used to investigate the microstructure of the deposits. For coated specimens with different phosphorus contents, there is one major peak at the 2θ position of 45° corresponding to the Ni (111) plane in the XRD profile. For the low phosphorus deposit, the diffraction peak is sharp due to the existence of a crystalline phase. However, a further widening of the diffraction peak with much reduced intensity at 2θ position of 45° indicates that the coating has a medium phosphorus content and the microstructure is composed of a mixture of amorphous and microcrystalline phases. When the phosphorus content in the coating is increased to 11.76 wt.%, the peak with reduced intensity becomes broader. The coating with a high phosphorus content has an amorphous structure. It is found from the above results that the microstructure changes from crystalline to amorphous state with an increase in the phosphorus content.

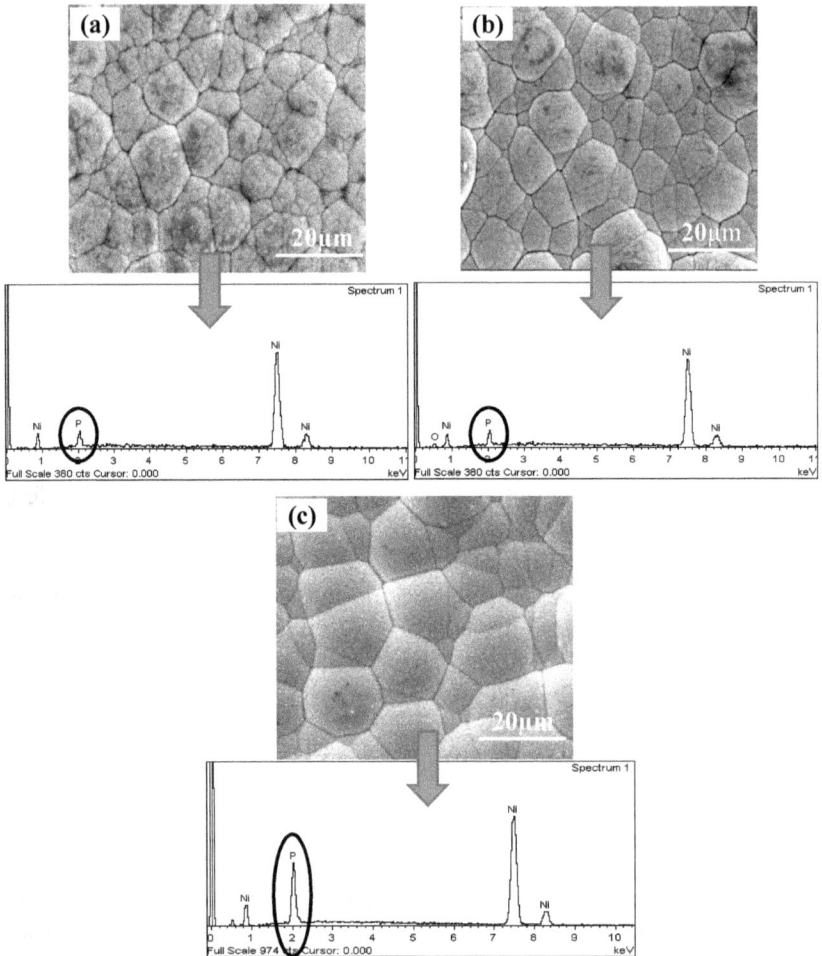

Figure 5-20 SEM images and EDS results of EN-P coating with different phosphorus content: (a) 3.25 wt.%; (b) 6.85 wt.%; (c) 11.76 wt.%

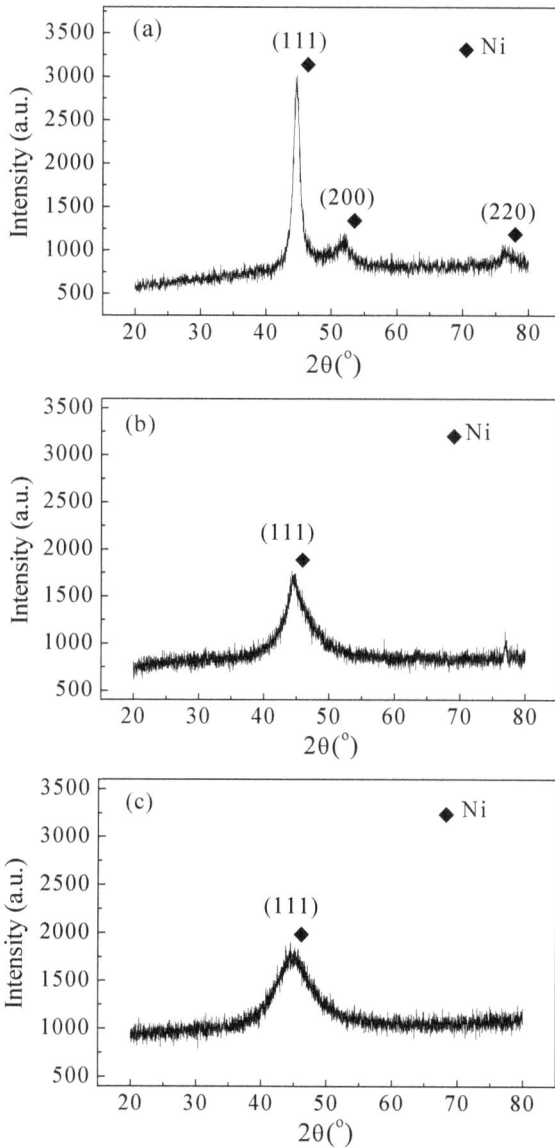

Figure 5-21 XRD patterns of EN-P coating with different phosphorus content: (a) 3.25 wt.%; (b) 6.85 wt.%; (c) 11.76 wt.%

5.2.3.3 Wear Performance of Ni-P Coating

The wear mass loss of the EN-P coating containing different phosphorus contents is summarized in Table 5-3. For the coating with 3.25 wt.% phosphorus, the mass loss is only 2.27 mg, while for the coating with 11.76 wt.% phosphorus the mass loss is 4.74 mg, showing that the wear mass loss of Ni-P coating increases with an increase in the phosphorus content. The hardness has the same trend in variation with the phosphorus content as the wear loss. Owing to the crystalline structure of the low phosphorus coating, the presence of phosphorus restricts the deformation of nickel crystal lattice, causing the improvement of the hardness. With an increase of the phosphorus content, the structure of the coating is gradually transitioned from crystalline to a disordered amorphous state, which weakens the block effect of phosphorus, leading to the reduction of the hardness. The hardness is not the only decisive factor affecting the wear resistance, but the decrease of hardness is one of the main reasons for the increased wear mass loss and the decreased wear resistance of the coating.

Table 5-3 Wear loss, hardness and surface roughness for EN-P coating with different phosphorus content

Phosphorus content (wt.%)	0	3.25	4.75	6.85	9.16	11.76
Wear loss (mg)	9.60	2.27	2.84	3.56	4.02	4.74
Hardness (HV)	378	665	613	588	557	516
Surface roughness (μm)	—	1.86	1.57	1.19	0.95	0.84

The coefficient of friction (COF) describes the ratio of the force of friction between two solid surfaces and the force pressing them together. It is one of the most importance parameters for the wear resistance of the material. The friction coefficient curves of the EN-P coating with different phosphorus contents are given in Figure 5-22.

The result is contrary to the change of the hardness. The rough surface and nodular structure explain the higher coefficient of friction for the low phosphorus coating. The smoother coating with the high phosphorus content results in a smaller friction coefficient. Because the hardness of the low phosphorus coating is greater than that of the high phosphorus coating, it possesses a stronger ability to resist adhesive wear. Vice versa, the high phosphorus coating is relatively soft, so it has a relatively weak wear resistance and more mass loss. This is the reason why the wear mass loss increases with an increase in the phosphorus content.

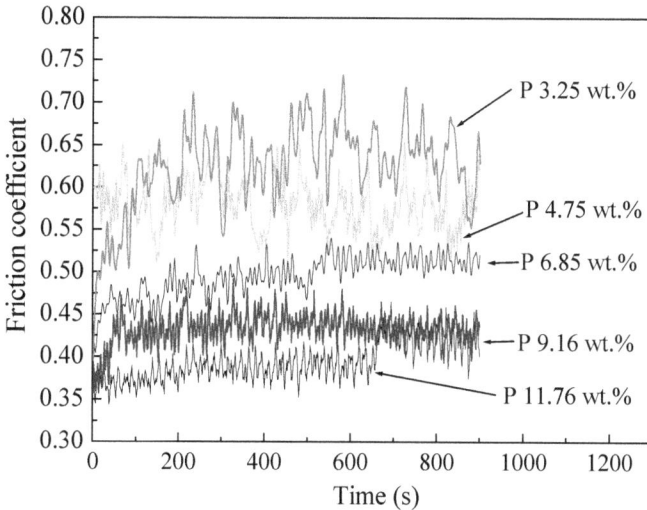

Figure 5-22 Friction coefficient of EN-P coating with different phosphorus content

The microhardness results reveal that the Ni-P coating has a superior resistance to indentation. Hence, it is concluded that the Ni-P coating notably enhances the hardness of Ti-6Al-4V.

5.2.4 Electroless Ni-P-MoS₂ Plating

5.2.4.1 Preparation of Ni-P-MoS₂ Composite Coating

The influences of the bath constituents (surfactants) and the process parameters (stirring and pH value) on the MoS_2 content are discussed to increase the MoS_2 content and homogeneity of in the composite coating.

1) pH

Figure 5-23 shows the effects of pH on the deposition rate and the MoS_2 content in the composite coating.

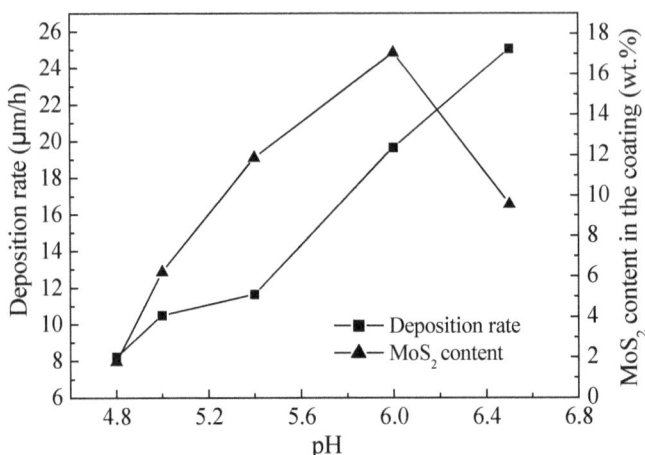

Figure 5-23 Effects of pH on deposition rate and MoS_2 concentration in EN-P-MoS_2 composite coating

The deposition rate increases with increasing pH. On the basis of the reactions (5.1)-(5.4), it can be seen that an increase in pH causes a decrease in the amount of reaction product H^+, which is beneficial to the reduction reaction of Ni^{2+}. This promotes the deposition of Ni. At pH which is less than 6, the MoS_2 content in the composite coating presents an increasing tendency. When pH is above 6, there is a noticeable drop in the MoS_2 content in the composite coating. The decrease of the H^+ concentration lowers the positively charged degree on the surface of the particles, blocking the adsorption and deposition of the MoS_2 particles. Nevertheless, the accelerated deposition rate reduces the probability of desorption, and accordingly the MoS_2 content in the composite coating still continues to increase. When the deposition of the Ni-P alloy and the desorption of the MoS_2 particles reach an equilibrium state, the probability of desorption increases, which causes a decrease of the MoS_2 content in the composite coating.

Figure 5-24 shows the surface morphologies and 3D contour graphs with the same magnification of Ni-P-MoS_2 composite coatings prepared at different pH values (a=4.8, b=5.0, c=5.5, d=6.0 and e=6.5).

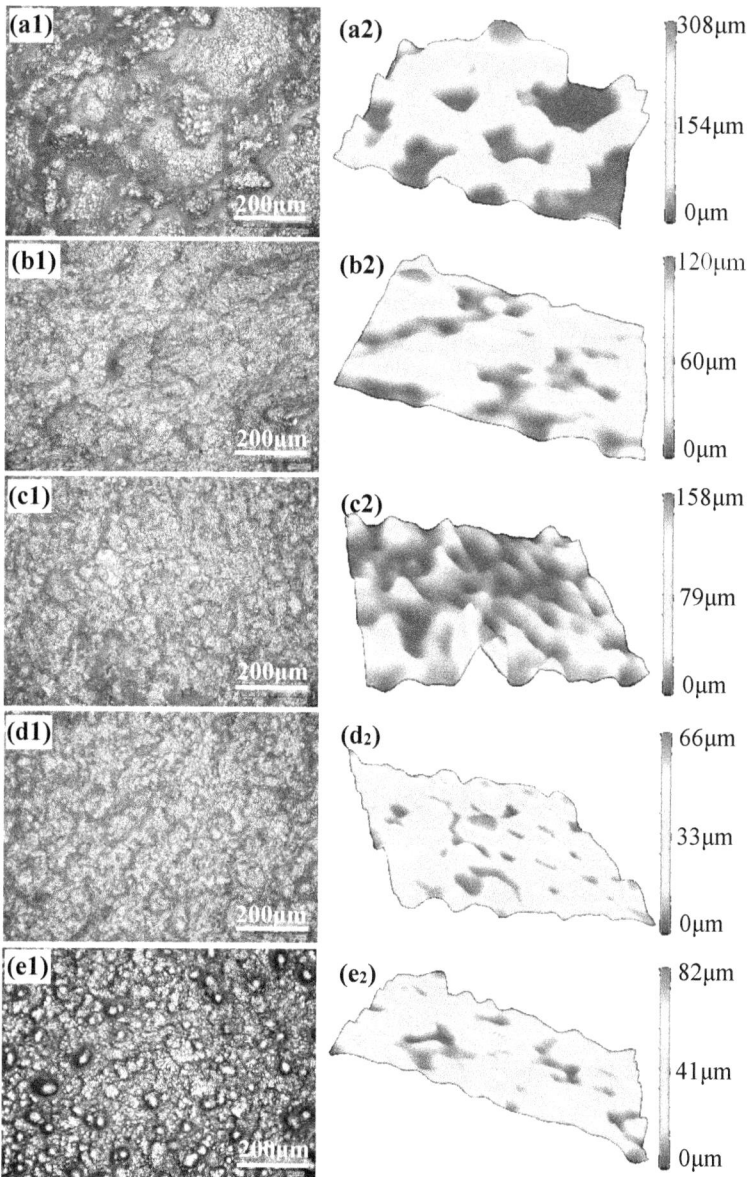

Figure 5-24 Surface morphologies and 3D contour graphs of EN-P-MoS$_2$ composite coating obtained at different pH Value: (a) pH=4.8; (b) pH=5.0; (c) pH=5.5; (d) pH=6.0; (e) pH=6.5

The convex portions in the 3D contour graph are the MoS$_2$ particles packed by a Ni-P deposit or absorbed on the surface of the Ni-P deposit. The height of the convex portion represents the coating thickness. The degree and uniformity of the undulating demonstrates the content and distribution uniformity of the MoS$_2$ particles in the Ni-P coating. With an increase in pH, the height of the convex portion in the Ni-P coating gradually increases, indicating that the deposition rate continuously increases. The quantities of the convex portion first increase and then decrease, indicating the MoS$_2$ content in the Ni-P coating has the same variation tendency. These observations are in accordance with the data in Figure 5-23. The change of pH affects the distribution of the MoS$_2$ particles in the Ni-P coating. At a pH of 6.0, the convex portions on the whole surface have a similar height and even distribution (Figure 5-24d). The results show that the MoS$_2$ particles are distributed homogeneously in the Ni-P coating under this condition.

2) MoS$_2$ concentration in bath

The deposition rate and MoS$_2$ content in the composite coating for the bath with varying concentration of MoS$_2$ in the bath are presented in Figure 5-25. The concentration of others is maintained constant.

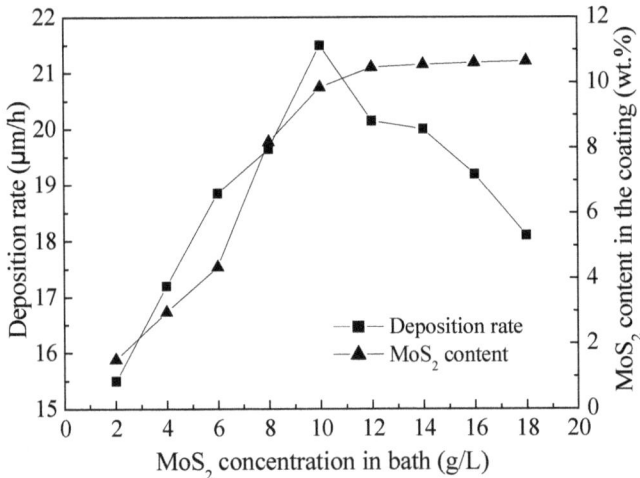

Figure 5-25 Effect of MoS$_2$ concentration on deposition rate and MoS$_2$ content of EN-P-MoS$_2$ composite coating

It is apparent that the deposition rate begins to increase gradually, and then declines as a function of increasing MoS_2 concentration in the bath. Simultaneously an increase in the concentration of MoS_2 in the bath results in an initial increase in the MoS_2 content in the composite coating followed by a plateau. An increase in the MoS_2 concentration in the bath raises the amount of MoS_2 particles suspended, improves the action of collision with the substrate surface, and leads to more active regions on the surface, hence, causing an increase in the deposition rate. Moreover, the collision probability of MoS_2 particles is heightened, and as a result more MoS_2 particles adsorb on the surface and further co-deposition of the MoS_2 particles in the Ni-P coating is promoted. When the MoS_2 concentration in the bath reaches 12 g/L, MoS_2 particle quantities incorporated in the composite coating have the highest value and then remain constant. This may be attributed to the reduction of the partial active regions covered by the MoS_2 particles incorporated in the coating, causing a decrease in the deposition rate.

3) Stirring method

Figure 5-26 shows the effect of stirring methods on the MoS_2 content and distribution in the Ni-P coating for a bath containing MoS_2 bath concentration (12g/L) at a temperature of 86°C and pH of 6.0.

Stirring can contribute significantly to uniform dispersion and the suspension of the MoS_2 particles in the bath and at the same time improves the collision probability on the surface. Compared with continuous mechanical stirring, continuous magnetic stirring obtains the coating with a much greater thickness and a much higher MoS_2 content. However, the distribution of the MoS_2 particles in the coating is uneven (Figure 5-26a). Continuous mechanical stirring produces a coating with evenly-distributed but less MoS_2 (Figure 5-26c). A magnetic rotator in continuous magnetic stirring constantly hit the bottom of the container. This makes the plating bath rapidly decompose, which explains the acceleration of the deposition rate and the increase of the MoS_2 content. These MoS_2 particles absorb on the surface of the Ni-P deposit and easily fall off. This kind of composite coating cannot improve wear resistance. Continuous mechanical stirring enables Ni^{2+}, $H_2PO_2^-$ and MoS_2 particles in the bath to remain uniform and continuously collide on the titanium surface, leading to an increase in the deposition rate and the desorption of MoS_2 particles. In terms of interval magnetic and mechanical stirring (stirring 20 min and stopping 5 min), the difference in the coating thickness between them is very small as a result of an almost identical deposition rate. However, the undulating degree showes that the coating prepared using interval magnetic stirring has a lower MoS$_2$ content and poor uniformity of MoS$_2$ particles (Figure 5-26b).

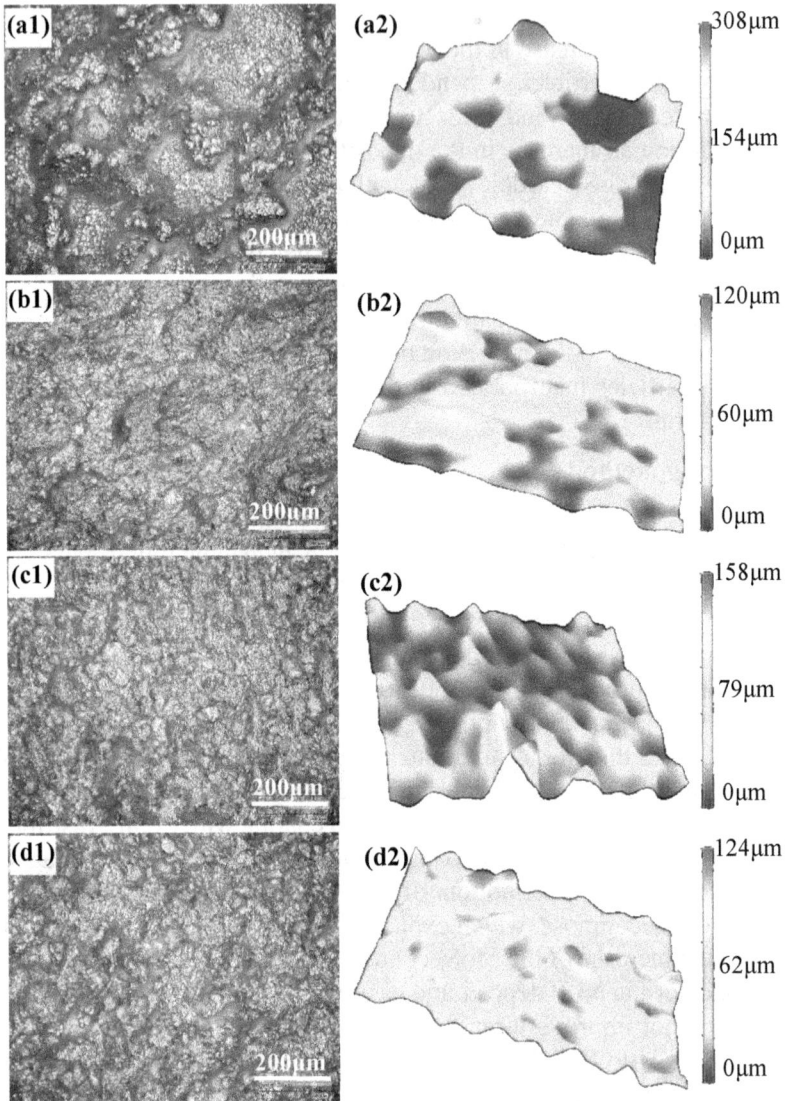

Figure 5-26 Surface morphologies and 3D contour graphs of EN-P-MoS$_2$ composite coating obtained using different stirring way: (a) continuous magnetic stirring; (b) interval magnetic stirring; (c) continuous mechanical stirring; (d) interval mechanical stirring

If interval mechanical stirring is applied, a coating with better homogeneity of MoS_2 particles is observed (Figure 5-25d). Interval stirring produces a composite coating with better quality in contrast with continuous stirring. This is mainly due to a better dispersion and suspension of Ni^{2+}, $H_2PO_2^-$ and MoS_2 particles in the bath, as well as an increase in the collision probability of MoS_2 particles on the surface of the titanium substrate. This stirring method not only speeds up the deposition rate, but also helps more MoS_2 particles stay on the substrate surface and incorporate into the growing Ni-P layer, which ensures the content and uniform distribution of MoS_2 particles in the coating.

Figure 5-27 shows the effect of the stirring rate on the MoS_2 content in the composite coating. As the stirring rate is increased from 0 to 350 rpm, the content increases firstly, peaks at a stirring rate of 200 rpm, and after that declines. Excessive stirring reinforces the impact strength of the bath to the substrate surface. This incurs desorption of the MoS_2 particles from the Ni-P coating and decreases the amount of the MoS_2 particles in the composite coating.

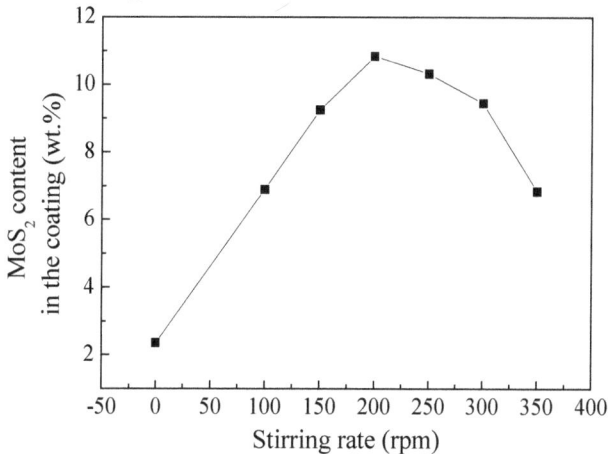

Figure 5-27 Effect of stirring rate on MoS_2 content of EN-P-MoS_2 composite coating

4) Surfactants

MoS_2 is a hydrophobic material. It floats on the surface of the bath and can not be dispersed or suspended in the bath. Surfactants are adopted for wetting treatment. The category and concentration of surfactants have strong influences on the dispersing effect of MoS_2 particles and the surface quality of the composite coating. Three different types of surfactant, cationic surfactant (hexadecyl trimethyl ammonium bromide), anionic surfactant (sodium lauryl sulphate), and mixed surfactant (cationic+anionic) are respectively added into a bath containing 8 g/L MoS_2. As can be seen from Table 5-4, using cationic surfactant obtains a high deposition rate and MoS_2 particle content in the composite coating. This is because a cationic surfactant after dissolution and ionization can absorb on the surface of the MoS_2 particles showing electronegativity, and wetting them. In addition, the steric hindrance effect of the surfactant can prevent agglomeration of the particles. These factors are good for the uniform dispersion and suspension of the MoS_2 particles, as well as the stability of the bath.

Table 5-4 Effects of surfactant type and concentration on deposition rate and MoS_2 content of EN-P-MoS_2 composite coating

Cationic (g/L)	Anionic (g/L)	Deposition rate (μm/h)	MoS_2 content in the coating (wt.%)
0.015	/	15.69	10.43
0.03	/	17.07	12.05
0.06	/	18.82	14.53
/	0.6	16.32	4.64
/	1.2	17.56	7.85
/	2.4	16.89	1.78
0.015	0.6	17.12	6.22
0.03	1.2	16.92	8.62
0.06	2.4	18.02	3.21

The above results are supported by the surface morphology of the Ni-P-MoS_2 composite coatings derived from the electroless Ni-P bath containing different types of surfactant are displayed in Figure 5-28.

The composite coating obtained by using a single cationic surfactant has more MoS_2 particles, but parts of which agglomerate. This can be inferred from the heterogeneous distribution of Mo and S elements in some areas (Figure 5-28a). When a single anionic surfactant substitutes a single cationic surfactant, agglomeration of MoS_2 particles is abated, but the amount of MoS_2 particles is also much less (Figure 5-28b). The combination use of a cationic and anionic surfactant is considered. The synergetic effect reduces the surface tension between the MoS_2 particles and the bath, fully wets and dispersedly suspends the MoS_2 particles in the bath, providing a favorable condition for the co-deposition of the Ni-P alloy and the particles. As a consequence, this enables a great deal of MoS_2 particles to be deposited with uniform and dispersive pattern in the Ni-P coating, which exhibits a nodular compact structure

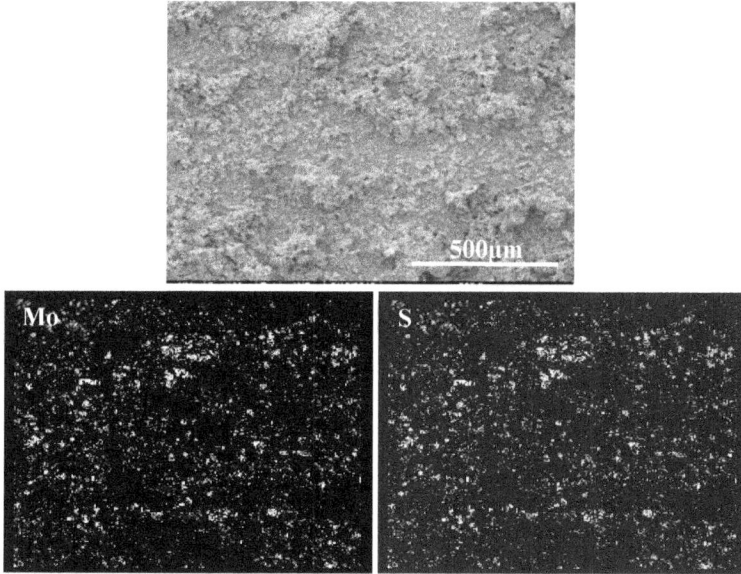

(a) cationic surfactant

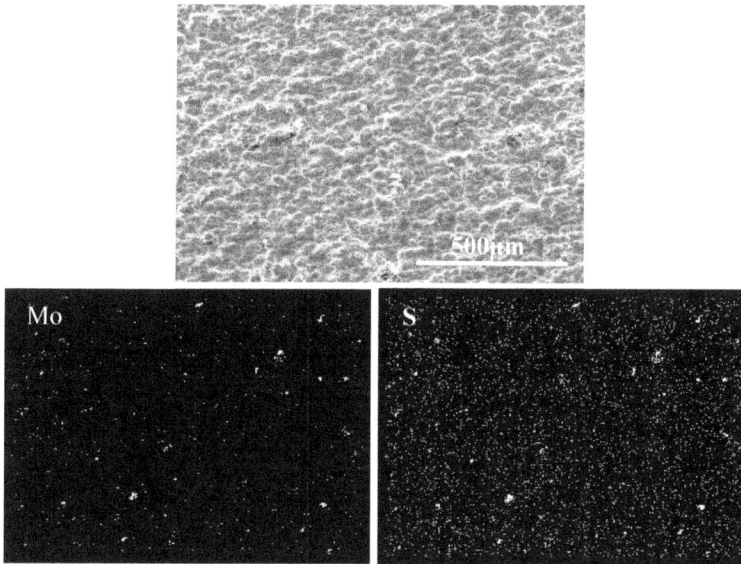

(b) anionic surfactant

Figure 5-28 Surface morphology and EDS analysis result of EN-P-MoS$_2$ composite coating obtained using different surfactants

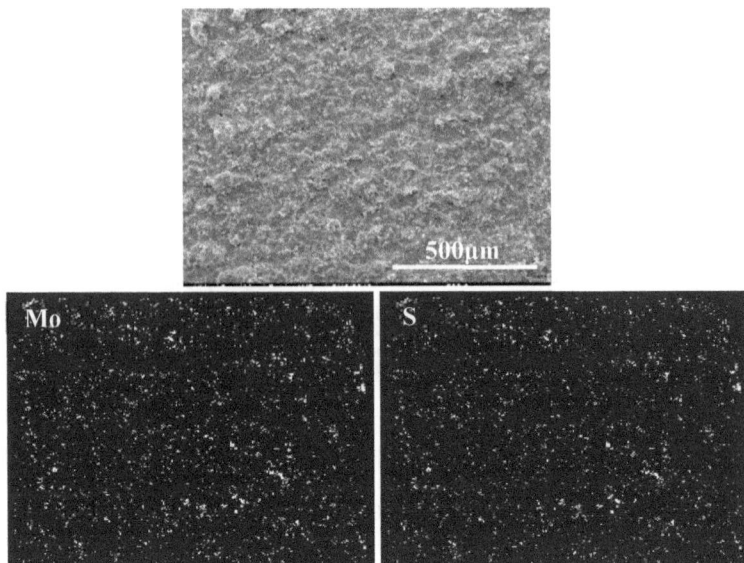

(c) mixed surfactant (cationic+anionic)

Figure 5-28 Surface morphology and EDS analysis result of EN-P-MoS$_2$ composite coating obtained using different surfactants (continued)

(Figure 5-28c). It can be drawn that the optimal option is 0.03 g/L of cationic surfactant and 1.2 g/L of anionic surfactant by taking into account the deposition rate and the MoS$_2$ distribution and content in the coating.

5.2.4.2 Co-deposition Mechanism

The co-deposition of particles in the coating has been discussed recently, so far three sorts of mechanisms have been proposed: mechanics, adsorption, and electrochemistry. In the mechanics mechanism, through external force the particles collide with the substrate surface and then stay on it, after which they are captured by a growing metal coating. In this theory, the particles are incorporated into the coating by a simple mechanics process. Thus, co-deposition is governed by a fluid mechanics factor and metal deposition rate. With regards to the adsorption mechanism, the prerequisite of co-deposition is that the particles can be adsorbed on the substrate under Van der Waals' force and are embedded in the growing metal after. It is pointed out in the electrochemistry mechanism that the key factor of particles deposited in the coating is the charge of the particle surface and field intensity formed in the particle/solution interface. Accordingly, charged particles are electrochemically absorbed on the surface by virtue of field intensity.

1) Bath zeta potential test

MoS_2 particles continuously collide with the substrate surface by the combined action of stirring and bath flowing force in the course of electroless plating. This leads to the incorporation of the MoS_2 particles in the Ni-P layer formed consecutively on the titanium surface, so that the co-deposition of Ni-P and MoS_2 can be implemented.

Figure 5-27 denotes that an increased stirring rate causes an increase in MoS_2 content, which coincides with the mechanics mechanism.

During the deposition process of MoS_2 particles, charged condition, charge density and types on the surface of MoS_2 particles can influence the particle migration in the electric field and interaction with the substrate surface, and then affect the distribution and co-deposition quantities of MoS_2 particles in the composite coating.

Zeta potential can be used to analyze the charge state on the surface of the composite particles. Figure 5-29a shows the variation of zeta potential with the MoS_2 concentration in the bath. Zeta potential is increased from 55.6 mV to 84.6 mV when the MoS_2 concentration in the bath is changed from 2 g/L to 16 g/L. After the MoS_2 concentration is more than 12 g/L, the increasing tendency of zeta potential slowes down. According to the electrochemical mechanism, a more positive zeta potential indicates a more positive charge on the particle surface and a higher co-deposition amount of Ni^{2+} and MoS_2 particles on the active areas, resulting in a higher MoS_2 content in the composite coating. The zeta potential as a function of the stirring rate is shown in Figure 5-29b. The zeta potential

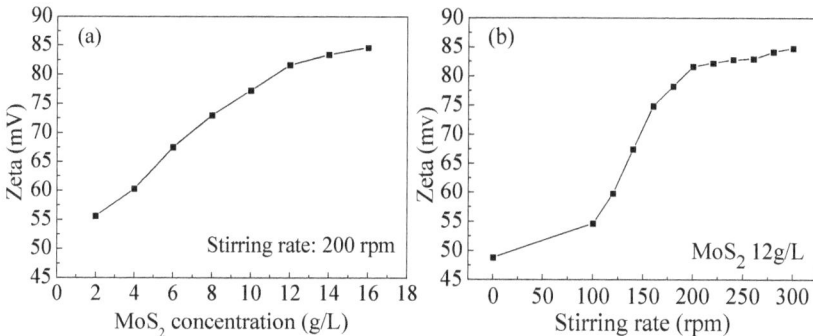

Figure 5-29 Effect of MoS_2 concentration (a) and stirring rate (b) in the bath on zeta potential

increases with an increase in the stirring rate. These results suggest that the mechanics mechanism and electrochemical mechanism simultaneously exist in the co-deposition of the Ni-P and MoS_2 particles and the mechanics effect plays a major role.

2) Activation of MoS_2 Particles in the Bath

MoS_2 particles may be activated in the electroless plating process, and nickel will deposit on the surface of the particles. The activation of the MoS_2 particles involves the formation of active spots, the adsorption of charge and the reduction of nickel ion to nickel on the surface of the MoS_2 particles. These spots covered by nickel deposit further become the catalytic activity centers of the electroless plating.

When the MoS_2 particles in the bath are fully activated, the MoS_2 particles are completely coated with a cellular Ni-P coating, and the EDS scanning results shows that the Ni and P elements are present on the surface of the MoS_2 particles (Figure 5-30). In this scenario, there may be two modes by which the MoS_2 particles are incorporated into the Ni-P coating:

a) The nickel ions on the surface of the particles are reduced to metal nickel, and when the particles reach the surface of the Ni-P deposit and are captured, the catalytic action of the newly-formed nickel enables the surface of the particles to be wrapped completely by the nickel layer;

b) The nickel ions on the surface of the particles are partly reduced to metal nickel, and the particles are imbedded in the Ni-P deposit by autocatalysis themselves.

Examination of a cross section of the composite coating shows that the MoS_2 particles are entirely trapped in the coating (Figure 5-31).

When the MoS_2 particles in the bath are partly activated, only the active portions are covered by the nickel deposit. The distribution of the P and Ni elements reveals that there are no Ni and P elements in some regions on the surface of the MoS_2 particles (Figure 5-32). Under this circumstance, the way in which the MoS_2 particles are embedded in the Ni-P deposit is similar to full activation, the difference is that parts of the MoS_2 particle surfaces are wrapped by the Ni-P deposit.

The MoS_2 particles which are inactivated in the bath are absorbed, but cannot be trapped in the Ni-P deposit. They easily fall off from the substrate. Ni and P elements are seldom detected on the surface of the MoS_2 particles (Figure 5-33).

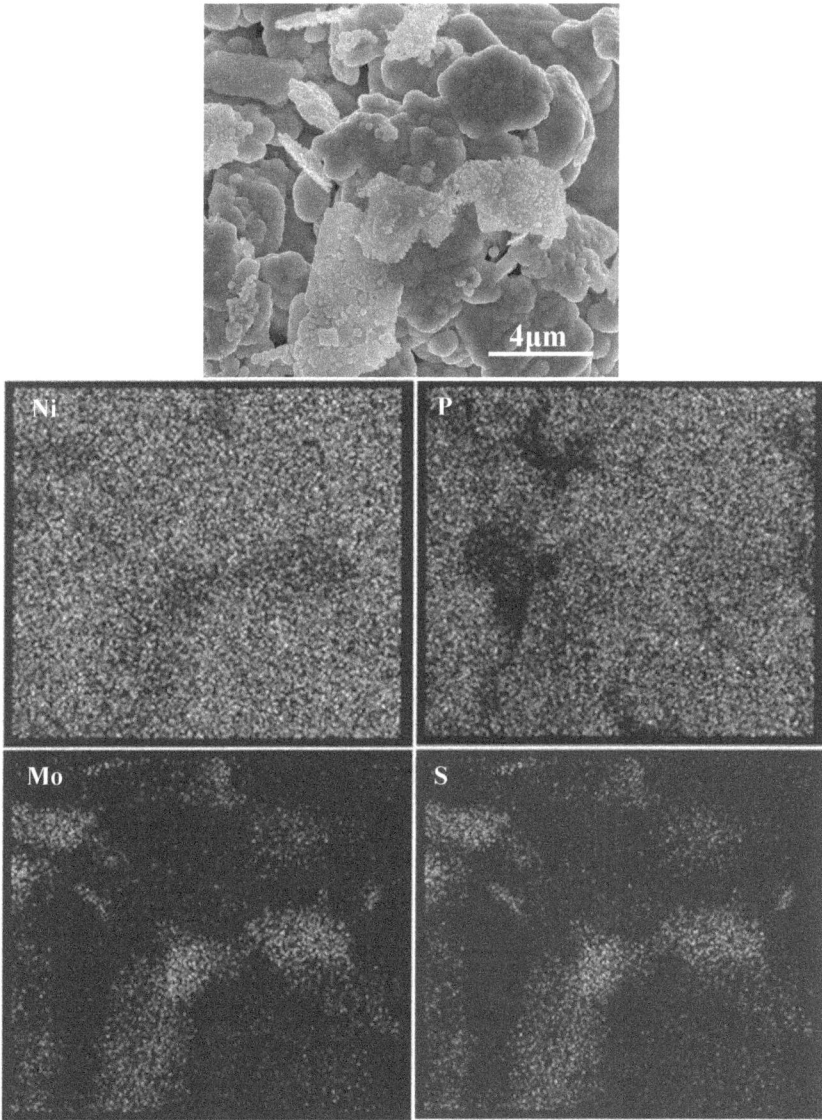

Figure 5-30 Surface morphology and EDS area scanning result of EN-P-MoS$_2$ composite coating with MoS$_2$ particles completely activated in the bath

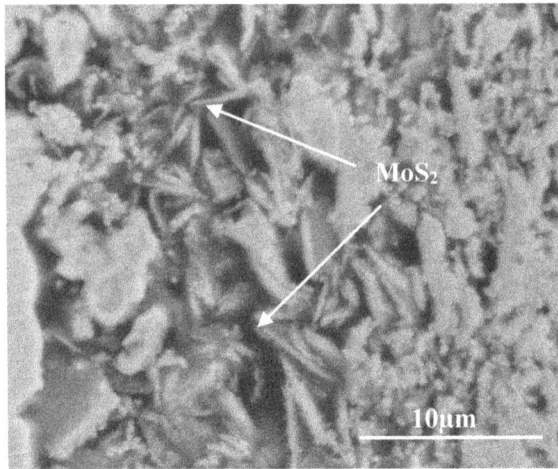

Figure 5-31 Cross section morphology of Ni-P-MoS$_2$ composite coating with MoS$_2$ particles completely activated in bath

The co-deposition process of Ni-P and MoS$_2$ particles goes through four stages:

(a) MoS$_2$ particles are uniformly dispersed in the bath and absorb the charge with the aid of surfactants;

(b) MoS$_2$ particles are delivered to the surface of the Ni-P deposit growing on titanium alloy by the impact force engendered by stirring;

(c) The mutual work of electrostatic attraction and mechanical collision promotes MoS$_2$ particles to be captured by the Ni-P deposit;

(d) The interfacial force between the MoS$_2$ particles on the surface of the Ni-P deposit and newly-formed nickel is produced in a high electric field in the Ni-P/bath interface. Depending on this force, fully activated

MoS$_2$ particles are completely packed in the Ni-P deposit and partly activated particles are incompletely covered by the Ni-P deposit. Inactivated particles are only adhered on the surface of the Ni-P deposit, and some particles are easily detached from the surface of the Ni-P deposit by gravity and bath scour.

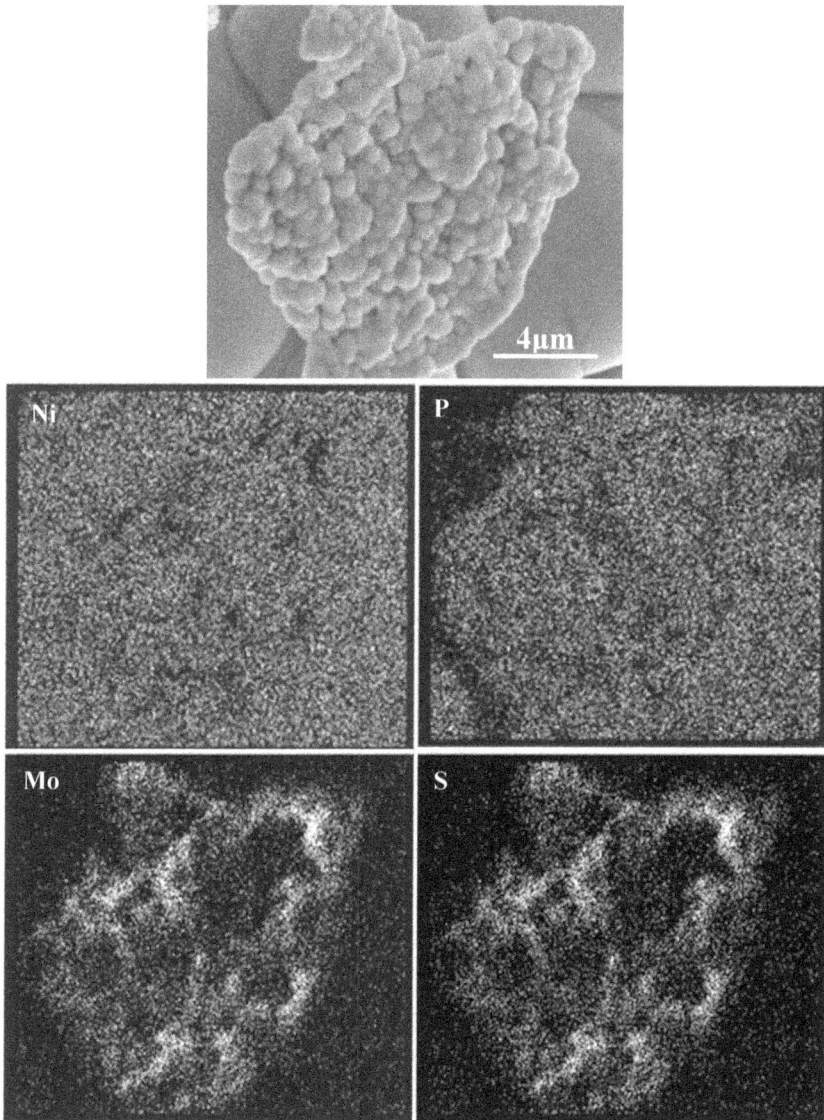

Figure 5-32 Surface morphology and EDS area scanning result of EN-P-MoS$_2$ composite coating with MoS$_2$ particles partly activated in bath

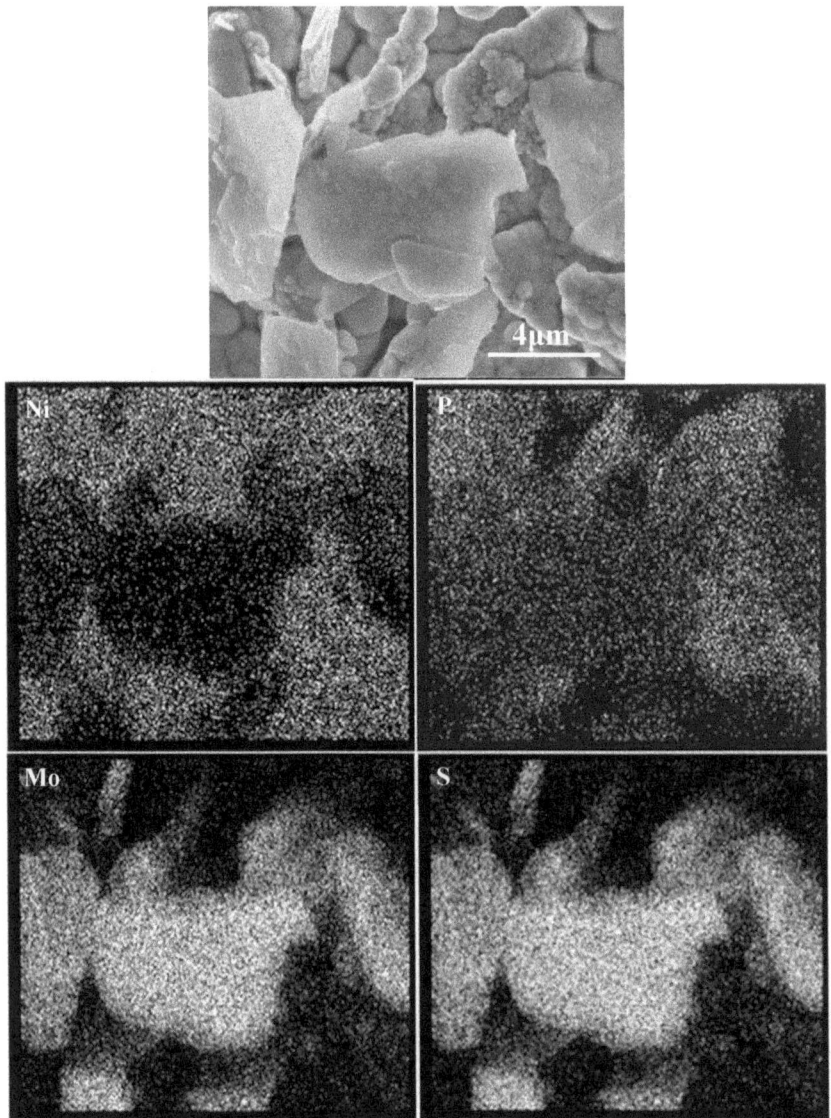

Figure 5-33 Surface morphology and EDS area scanning result of EN-P-MoS$_2$ composite coating with MoS$_2$ particles inactivated in bath

5.2.4.3 Wear Performance of Ni-P-MoS₂ Coating

1) Effect of MoS₂ content in the composite coating

The dynamic friction coefficient curves of the composite coating with varying MoS₂ content are displayed in Figure 5-34.

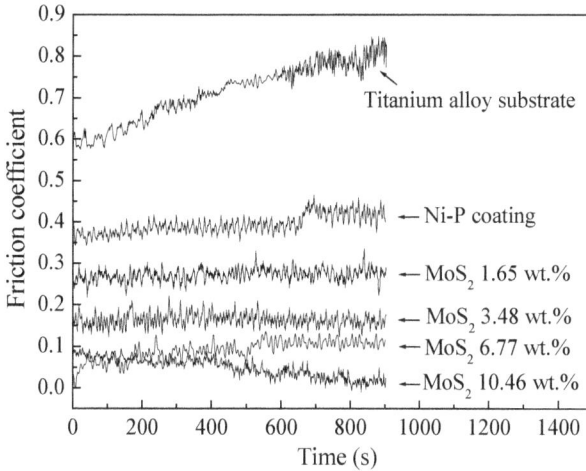

Figure 5-34 Effect of MoS₂ content on friction coefficient of EN-P-MoS₂ composite coating

As can be seen from the figure, the coefficient of friction of Ti-6Al-4V is the highest and a gradual increase from 0.6 to 0.9 is found with an increase in wear time. The EN-P coating with a phosphorus content of 11.76 wt.% possesses a lower friction coefficient by contrast withTi-6Al-4V substrate, fluctuating between 0.35 and 0.40. For the EN-P-MoS₂ composite coating, the friction coefficient declines as the MoS₂ content in the composite coating goes up. The average friction coefficient of the composite coating with MoS₂ content of 1.65 wt.%, 3.48 wt.%, 6.77 wt.% and 10.46 wt.% are respectively 0.27, 0.17, 0.09 and 0.05, which are far lower than those of the titanium substrate and the Ni-P coating. Figure 5-35 shows the wear loss of the EN-P-MoS₂ composite coating with 1.65 wt.%, 3.48 wt.%, 6.77 wt.% and 10.46 wt.% MoS₂. The composite coating with a higher MoS₂ content has much less wear loss.

It can be inferred from the results that the composite coating has excellent self-lubrication and anti-friction properties, and a better anti-friction effect can be achieved for the composite coating with a higher MoS₂ content.

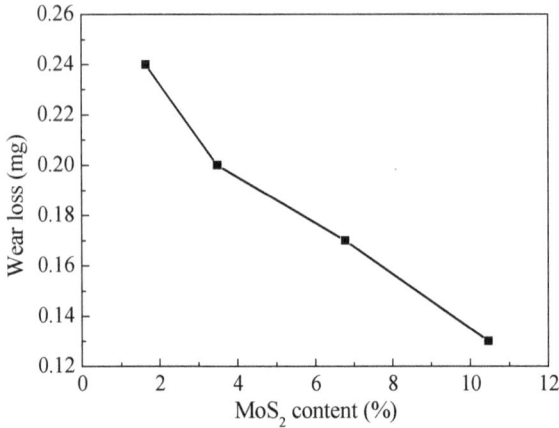

Figure 5-35 Effect of MoS₂ content on wear loss of EN-P-MoS₂ composite coating

2) Effect of heat treatment temperature

Figure 5-36 depicts the influence of the temperature of heat treatment on the friction coefficient of the composite coating. The friction coefficient presents a decreasing tendency when the temperature of heat treatment is increased, all of which are lower than that of the composite coating in as-deposited condition.

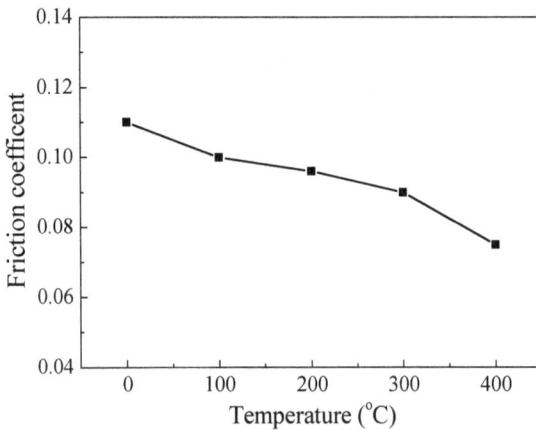

Figure 5-36 Effect of temperature of heat treatment on friction coefficient of EN-P-MoS₂ composite coating (MoS₂ wt.%=6.77)

The XRD spectra of the composite coating in as-deposited condition and after heat treatment at 400°C are given in Figure 5-37. There is one major broad-peak at the 2θ position of 45° corresponding to Ni (111) plane in XRD profile of the composite coating as-deposited, in addition, eight sharp peaks corresponding to MoS_2 are observed. After heat treatment at 400°C, the diffraction peaks of Ni_3P phase appear in XRD patterns. These phases disperse in the Ni-P deposit, acting as a strengthening phase. Along with the MoS_2 particles, they can significantly improve the hardness and enhance the plastic deformation resistance of the composite coating.

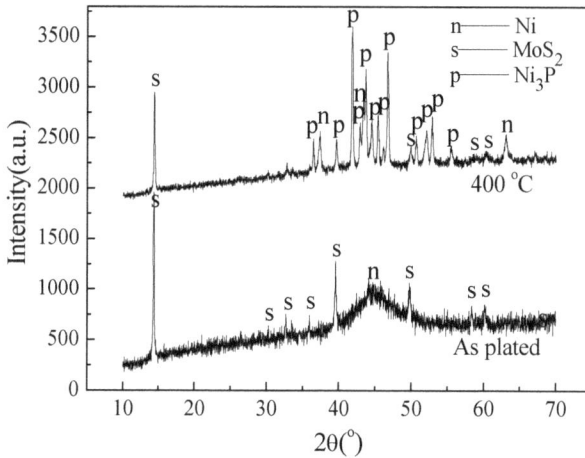

Figure 5-37 XRD patterns of EN-P-MoS_2 composite coatings as-deposited and after heat treatment at 400°C (MoS_2 wt.%=6.77)

3) Wear behavior

MoS_2 particles with a low friction coefficient possess good anti-friction properties. The metallographic images of the worn composite coating surfaces after wear examined by optical microscopy are represented in Figure 5-38. In these figures, the black parts are the MoS_2 particles, and the bright regions show the morphology of occlusion and spalling after the wear test. Some small wear debris stick to the wear tracks, moreover the phenomena of occlusion and spalling occur on the worn surface. From Figures 5-38(a) and (b), a coexistent characteristic of abrasive and adhesive wear is found, suggesting that the composite coating with a low MoS_2 content has a relatively poor self-lubricating property. In terms of the composite coating with a high MoS_2 content, a large amount of MoS_2 particles are incorporated in the composite coating, which are softened under friction movement, and then uniformly distributed on the microcosmic uneven friction surface, ultimately forming a layer of uniform solid lubricating film in the contact surface. This film has a good solid self lubrication performance, which prevents the occurrence of adhesive wear (Figure 5-38c).

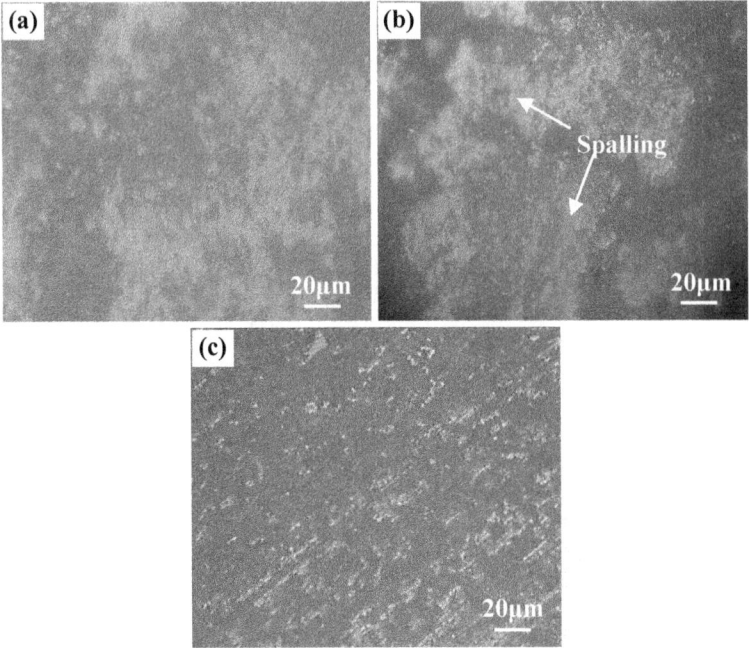

Figure 5-38 Wear morphology of electroless Ni-P-MoS$_2$ composite coating with different MoS$_2$ content: (a) 3.48 wt%; (b) 6.77 wt%; (c) 10.46 wt%

References

Ahmad, Z., 2006. Principles of corrosion engineering and corrosion control. Elsevier, Boston, MA, 537-545.

Armco Steel Corporation, 1973. Chemical milling process and bath therefore. The United Kingdoms, 1304043.

ASTM 32-16, 2016. Standard test method for cavitation erosion using vibratory apparatus.

Aviation Manufacturing Engineering Handbook (AMEH) Editorial Board, 1993. Aviation Manufacturing Engineering Handbook - Special processing, 600-645. (Chinese)

AZO Materials, 2002. Titanium–applications. http://www.azom.com/article.aspx?ArticleID=1297.

Bansal, D. G., O. L. Eryilmaz and P. J. Blau, 2011. Surface engineering to improve the durability and lubricity of Ti-6Al-4V alloy. Wear, **271**(9-10): 2006-2015.

Bardal, E., 2004. Corrosion and protection. Springer, London, 152-154, 170.

Barik, R. C., J. A. Wharton, R. J. K. Wood and K. R. Stokes, 2009. Electromechanical interactions during erosion-corrosion. Wear, **267**(2): 1900-1908.

Bloyce, A, P. Y. Qi, H. Dong and T. Bell, 1998. Surface modification of titanium alloys for combined improvements in corrosion and wear resistance. Surface and Coatings Technology, **107**(2-3): 125-132.

Boyer R. R., 2010. Attributes, characteristics, and applications of titanium and its alloys. JOM, **62**(5): 35-43.

Bregliozzi, G., A. D. Schino, S. I. U. Ahmed, J. M. Kenny and H. Haefke, 2005. Cavitation wear behavior of austenitic stainless steels with different grain sizes. Wear, **258**(1-4): 503-510.

Brunatto, S. F., A. N. Allenstein and C. L. M. Allenstein and J. A. Buschinelli, 2012. Cavitation erosion behavior of niobium. Wear, **274-275**: 220-228.

Budzynski, P., A. Youssef and J. Sielanko, 2006. Surface modification of Ti–6Al–4V alloy by nitrogen ion implantation. Wear, **261**(11-12): 1271-1276.

Cakir, O., 2008. Chemical etching of aluminium. Journal of Material Processing and Technology, **199**(1-3): 337-340.

Cakir, O., H. Temel and M. Kiyak., 2005. Chemical etching of Cu-ETP copper. Journal of Materials Processing Technology, **162-163**: 275-279.

Chiu, K.Y., F. T. Cheng and H. C. Man, 2005. Evolution of surface roughness of some metallic materials in cavitation erosion. Ultrasonics, **43**: 713-716.

Corrosionpedia, 2016. Diffusion Coating. https://www.corrosionpedia.com/definition/394/diffusion-coating.

Dadvand, N, 2002. Investigation of the corrosion behavior of electroless nickel-boron and nickel-phosphorus coatings in basic solutions. Dalhousie University, Halifax, 79-85.

Davim, J. P., 2012. Wear of advanced materials. Wiley, London, 129-131.

Diamanti, M. V. and M. P. Pedeferri, 2007. Effect of anodic oxidation parameters on the titanium oxides formation. Corrosion Science, **49**(2): 939-948.

Ding, L. X. and Y. M. Wang, 2006. Wear resistant coatings and properties of titanium and its alloys. Northeastern University Press, Shenyang, 50-53. (Chinese)

Donachie, M. J., 2000. Titanium - A technical guide. ASM International, Ohio, 1-3.

Fei, C., Z. Hai, C. Chen and Y. J. Xia, 2009. Study on the tribological performance of ceramic coatings on titanium alloy surfaces obtained through microarc oxidation. Progress in Organic Coatings, **64**(2-3): 264-267.

Fernández-Domene R. M., B. Blasco-Tamarit, D. M. García-García and J. García-Antón, 2011. Cavitation corrosion and repassivation kinetics of titanium in a heavy brine LiBr solution evaluated by using electrochemical techniques and Confocal Laser Scanning Microscopy. Electrochimica Acta, **58**(30): 264-275.

Fernández-Domene, R. M., E. Blasco-Tamarit, D. M. García-García and J. García-Antón, 2010. Repassivation of the damage generated by cavitation on UNS N08031 in a LiBr solution by means of electrochemical techniques and Confocal Laser Scanning Microscopy. Corrosion Science, **52**(10): 3453-3464.

García-García, D. M., J. García-Antón and A. Igual-Muñoz, 2008. Influence of cavitation on the passive behavior of duplex stainless steels in aqueous LiBr solutions. Corrosion Science, **50**(9): 2560-2571.

García-García, D. M., J. García-Antón, A. Igual-Muñoz and E. Blasco-Tamarit, 2006. Effect of cavitation on the corrosion behavior of welded and non-welded duplex stainless steel in aqueous LiBr solutions. Corrosion Science, **48**(9): 2380-2405.

Guan, X., 2010. Cavitation erosion behavior of titanium and Ti-6Al-4V alloy. Tianjin University, Tianjin. (Chinese)

Guo, C., J. S. Zhou, J. M. Chen, J. R. Zhao, Y. J. Yu and H. D. Zhou, 2010. Improvement of the oxidation and wear resistance of pure Ti by laser cladding at elevated temperature. Surface and Coatings Technology, **205**(7): 2142-2151.

Gurrappa, I., 2003. Characterization of titanium alloy Ti-6Al-4V for chemical, marine and industrial applications. Materials Characterization, **51**(2-3): 131-139.

Harris, W. T., 1976. Chemical milling: the technology of cutting materials by etching. Clarendon Press, Oxford, 43-72.

Hu, G., 2011. The study on dissolution behavior of corrosion processing and fatigue performance after processing for titanium alloy. Nanchang Hangkong University, Nanchang, China. (Chinese)

Huang, B. Y., C. G. Li, L. K. Shi, Q. Z. Qiu and T. Y. Zuo, 2009. Handbook of non-ferrous metal materials. Chemical Industry Press, Beijing, 513-740. (Chinese)

Huang, J. Z. and Y. Zuo, 2003. Corrosion resistance and corrosion data of materials. Chemical Industry Press, Beijing, 320-325. (Chinese)

IBC Coatings Technologies INC., 2016 Micro Arc Oxidation-CeraTough™. http://www.ibccoatings.com/ plasma-electrolytic-oxidation-peo-ceratough.

International Titanium Association, 2011. Titanium-the infinite choice. www.titanium.org.

Jiang, X. X., S. Z. Li and S. Li, 2003. Corrosive wear of metals. Chemical Industry Press, Beijing, 196p. (Chinese)

Jin, H., G. S. Lv, G. Y. Cai and Z. Ma, 2005. Abrasivity study of different types of electroless nickel deposits. Electroplating & Finishing, **24**(2): 4-6.

Jin, L. And D. Li, 1989. Chemical milling and electrochemical machining. Rare Metal Materials and Engineering, (2): 66-71. (Chinese)

Kim, Y. S., H. I. Kim, J. H. Cho, H. K. Seo, G. S. Kim, S. G. Ansari, G. Khang, J. J. Senkevich and H. S. Shin, 2006. Electrochemical deposition of copper and ruthenium on titanium. Electrochimica Acta, **51**(25): 5445-5451.

Knapp, R. T., 1955. Recent investigations of the mechanics of cavitation and cavitation damage. Transactions of the ASME, **75**(8): 1045-1054.

Kwok, C. T., H. C. Man, and L. K. Leung, 1997. Effect of temperature, pH and sulphide on the cavitation erosion behaviour of super duplex stainless steel. Wear, **211**(1): 84-93.

Kwok, C. T., F. T. Cheng and H. C. Man, 2000. Synergistic effect of cavitation erosion and corrosion of various engineering alloys in 3.5% NaCl solution. Materials Science and Engineering A, **290**: 145-154.

Laser Cladding Technology Ltd, 2016. Laser cladding process. http://www.lasercladding.co.uk/Laser-Cladding-Process.aspx.

Leyens, C. and M. Peters, 2003. Titanium and titanium alloys-fundamentals and applications. WILEY-VCH GmbH & Co.kGaA, Weinheim, Germany, 1-4, 9, 334-349.

Li, D. L., G. C. Zou, G. T. Zheng and X. Y. Yong, 2012. Effect of Microstructure on surface layer mechanical properties of stainless steel under cavitation. Chinese Journal of Materials Research, 26(3): 274-278. (Chinese)

Liang, J., 2010. The study on technology and mechanism of precise corrosion processing of titanium alloy. Nanchang Hangkong University, Nanchang, China. (Chinese)

Lim, P. Y., P. L. She and H. C. Shih, 2006. Microstructure effect on microtopography of chemically etched α+β Ti alloys. Applied Surface Science, 253(2): 449-458.

Lin, C. and L. C. Zhao, 2013. Electroless Ni-P wear-resistant coating on TC4 titanium alloy. Rare Metal Materials and Engineering, 42(3): 507-512. (Chinese)

Lin, C. and N. Du, 2014. Selection and design of titanium alloy. Chemical Industry Press, Beijing, 1-11, 58-89, 97-128. (Chinese)

Lin, C. and X. P. Hong, 2011. Investigation of corrosion processing for Ti-6Al-4V in hydrochloric-nitric acid system. Proceedings of the ASME 2011 International Manufacturing Science & Engineering Conference, June 13-17, Corvallis, Oregon, USA. DOI: 10.1115/MSEC2011-50111.

Lin, C., F. Liu, Q. Zhao, Q. J. Wen and N. Du, 2008. Influencing factors of rate and surface quality of corrosion processing for TC4. Journal of Aeronautical Materials, 28(5): 50-54. (Chinese)

Lin, C., G. Hu, J. Liang, Q. Zhao, N. Du and L. Q. Wang, 2010. Dissolution behavior of corrosion processing for TC1 and TC4 titanium alloy. Journal of Aeronautical Materials, 30(6): 37-44. (Chinese)

Lin, C., L. C. Zhao, H. F. Zhao and Q. Y. Wu, 2013a. Effect of phosphorous content on microstructure and wear resistance of electroless Ni-P alloy coatings on titanium alloy. Journal of Materials Protection, 46(1): 5-7. (Chinese)

Lin, C., N. Dadvand, Z. Farhat and G. J. Kipouros, 2013b. Electroless nickel phosphorus plating on carbon steel. Materials Science and Technology (MS&T) 2013, October 27-31, Montreal, Quebec, Canada.

Lin, C., N. Du, G. Hu and Y. F. Zhang, 2016a. Corrosion processing dissolution characteristics of Ti-6Al-4V in hydrofluoric-nitric acid system. Rare Metal Materials and Engineering, 45(10): 2628-2635. (Chinese)

Lin, C., Q. Y. Wu, and X. B. Zhao, 2015a. Electroless Ni-P-MoS$_2$ composite coating on TC4 titanium alloy. Corrosion & Protection, **36**(5): 412-418. (Chinese)

Lin, C., Q. Zhao and Q. J. Wen, 2015b. Corrosion processing for TC1 titanium alloy and its effect on substrate properties. Journal of Materials Engineering, **43**(7): 48-55. (Chinese)

Lin, C., X. B. Zhao and Y. F. Zhang, 2016b. Research progress on cavitation-corrosion of metallic materials. Journal of Chinese Society for Corrosion and Protection, **36**(1): 11-18. (Chinese)

Liu, F., 2008. The study on chemical milling process and performance of titanium alloy. Nanchang Hangkong University, Nanchang, China. (Chinese)

Liu, Q., B. Li and X. Z. Chen, 2012. Study on a Cr-Co-Ni-Mn austenitic stainless steel for cavitation erosion. Journal of Functional Materials, **43**(5): 673-676. (Chinese)

Liu, S. H. and D. L. Chen, 2009. Mechanics mechanism of duplex steel cavitation damage. Acta Metallurgica Sinica, **45**(5): 519-526. (Chinese)

Liu, W., Y. G. Zheng, C. S. Liu, Z. M. Yao and W. Ke, 2003. Cavitation erosion behavior of Cr–Mn–N stainless steels in comparison with 0Cr13Ni5Mo stainless steel. Wear, **254**(7-8): 713-722.

Lütjering, G. and J. C. Williams, 2007. Titanium, Springer, New York, 15-18.

Madina, V. and I. Azkarate, 2009. Compatibility of materials with hydrogen. Particular case: Hydrogen embrittlement of titanium alloys. International Journal of Hydrogen Energy, **34**(14): 5976-5980.

Mahmoud, S. S., 2009. Electroless deposition of nickel and copper on titanium substrates: Characterization and application. Journal of Alloys and Compounds, **472**(1-2): 595-601.

Mandani, F., H. Ettouney and H. El-Dessouky, 2000. LiBr-H$_2$O absorption heat pump for single-effect evaporation desalination process. Desalination, **128**(2): 161-176.

McDonnell Douglas Corporation, 1974. Chemical milling of titanium, refractory metals and their alloys. The United States, 3788914.

McDonnell Douglas Corporation, 1977. Chemical-milling of titanium and refractory metals. The United States, 4116755.

Metalpedia, 2017. Titanium resources, reserves and production. http://metalpedia.asianmetal.com/metal/titanium/resources&producti on.shtml.

Mochizuki, H., M. Yokota and S. Hattori, 2007. Effects of materials and solution temperatures on cavitation erosion of pure titanium and titanium alloy in seawater. Wear, **262**(5-6): 522-528.

Mogoda, A. and Y. Ahmad, 2004. Corrosion behavior of Ti-6Al-4V alloy in concentrated hydrochloric and sulphuric acid. Journal of Applied Electrochemistry, **34**(9): 873-878.

Mudali, U. K. and Raj, B., 2009. Corrosion science and technology: mechanism, mitigation and monitoring. Narosa Publishing House, New Delhi, India, 194-195, 199.

Neville, A. and B. A. B. McDougall, 2001. Erosion- and cavitation-corrosion of titanium and its alloys. Wear, **250**(1-12): 726-735.

Niu, W. and R. L. Sun, 2006. Research progress and development on laser cladding on titanium alloys surface. Materials Review, **20**(7): 58-60, 68. (Chinese)

North American Aviation, Inc., 1956. Process of chemically milling structural shapes and resultant article. The United States, 2739047.

Osterman, A., B. Bachertb, B. Siroka and M. Dular, 2009. Time dependant measurements of cavitation damage. Wear, **266**(9-10): 945-951.

Park, M. C., K. N. Kim, G. S. Shin, J. Y. Yun, M.H. Shin and S.J. Kim, 2013. Effects of Ni and Mn on the cavitation erosion resistance of Fe–Cr–C–Ni/Mn austenitic alloys. Tribology Letters, **52**(3): 477-484.

Patella, R. F., J. L. Rebouda and A. Archer, 2000. Cavitation damage measurement by 3D laser profilometry. Wear, **246**(1-2): 59-67.

Royal Society of Chemistry (RSC), 2016. Periodic table – titanium. http://www.rsc.org/periodic-table/element/22/titanium.

Sarma, B., N. M. Tikekar and K. S. R. Chandran, 2012. Kinetics of growth of superhard boride layers during solid state diffusion of boron into titanium. Ceramics International, **38**(8): 6795-6805.

Say, W. C. and Y. Y. Tsa, 2004. Surface characterization of cast Ti-6Al-4V in hydrofluoric nitric pickling solutions. Surface and Coatings Technology, **176**(3): 337-343.

Seiji, B. and I. Yukari, 2006. Surface modification of titanium by etching in concentrated sulfuric acid. Dental Materials, **22**(12): 1115-1120.

Shen, L. G., 1984. Chemical milling. National defence industry press, Beijing, 1-3, 6-7, 44-49.

Simka, W., A. Sadkowski, M. Warczak, A. Iwaniak, G. Dercz, J. Michalska and A. Maciej, 2011. Characterization of passive films formed on titanium during anodic oxidation. Electrochimica Acta, **56**(24): 8962-8968.

Takasaki, A. and Y. Furuya, 1996. Hydrogen evolution from chemically etched titanium aluminides. Journal of Alloys and Compounds, **243**(1-2): 167-172.

The chemical Engineer, 2012. Glauber-the chemical engineer. www.thechemicalengineer.com./~/media/Documents/TCE/Articles/2 012/851/851cewctw.pdf.

Tian, Y. S., C. Z. Chen and Q. H. Huo, 2005. Research progress on laser surface modification of titanium alloys. Applied Surface Science, **242**(1-2): 177-184.

Total Materia, 2005. Physical properties of titanium and its alloys. http://www.totalmateria.com/Article122.htm.

Wang Z. Y. and J. H. Zhu, 2003. Effect of phase transformation on cavitation erosion resistance of some ferrous alloys. Materials Science and Engineering A, **358**(1-2): 273-278.

Wang, B. C. and J. H. Zhu, 2008. Influence of unltrasonic cavitation on passive film of stainless steel. Ultrasonics Sonochemistry, **15**(3): 239-243.

Wang, Y. M., B. L. Jiang, T. Q. Lei and L. X. Guo, 2006. Microarc oxidation coatings formed on Ti-6Al4V in Na_2SiO_3 system solution: Microstructure, mechanical and tribological properties. Surface and Coatings Technology, **201**(1-2): 82-89.

Wang, Y. M., B. L. Jiang, T. Q. Lei, D. C. Jia and Y. Zhou, 2003. Research development of wear-resistance coatings on Ti6Al4V alloy surface. Materials Engineering, (3): 38-43. (Chinese)

Wang, Z. Y. and J. H. Zhu, 2003. Effect of phase transformation on cavitation erosion resistance of some ferrous alloys. Materials Science and Engineering A, **358**(1-2): 273-278.

Wikipedia, 2016. Chemical milling. https: //en.wikipedia.org/wiki/ Chemical_milling.

Wikipedia, 2017. Titanium alloy. https://en.wikipedia.Org/wiki/Titanium_alloy.

Wikipedia, 2017. Titanium. https://en.wikipedia.org/wiki/Titanium.

Wu, Q. Y., 2014. Investigation of electroless Ni-P plating on carbon fiber and electroless $Ni-P-MoS_2$ composite plating on titanium alloy. Nanchang Hangkong University, Nanchang, China. (Chinese)

Yang, D., 2008. Metal etching technology. National Defence Industry Press, Beijing, 1-4, 20-22, 143-147. (Chinese)

Yong, X. Y., J. Ji, Y. Q. Zhang, D. L. Li and Z. J. Zhang, 2011. AFM morphology and cavitation corrosion process of austenitic stainless steel by cavitation. Corrosion Science and Protection Technology, **23**(2): 116-120. (Chinese)

Zhang, X. Y., Y. Q. Zhao and C. G. Bai, 2005. Titanium alloy and its application. Beijing Chemical Industry Press, Beijing, 21-23.

Zhang, Y. F., 2015a. The study on cavitation corrosion behavior of Ti-6Al-4V alloy in lithium bromide solution. Nanchang Hangkong University, Nanchang. (Chinese)

Zhang, Y. F., C. Lin, N. Du and X. B. Zhao, 2015b. Cavitation corrosion behavior of TC4 titanium alloy in lithium bromide solution. Corrosion & Protection, **36**(6): 11-19. (Chinese)

Zhao, L. C., 2012. Investigation of electroless plating Ni-P, Ni-P-MoS$_2$ composite coating and their wear performance on the surface of titanium alloy. Nanchang Hangkong University, Nanchang, China. (Chinese)

Zhao, S. P., 2003. Titanium alloy and its surface treatment. Harbin Institute of Technology Press, Harbin, China, 112-116, 118-119. (Chinese)

Zhao, X. B., 2016. Research on behavior at initial stage and influencing factors of cavitation corrosion for Ti-6Al-4V alloy in LiBr solution. Nanchang Hangkong University, Nanchang, China. (Chinese)

Zheng, Y. G., S. Z. Luo and W. Ke, 2008. Cavitation erosion-corrosion behavior of CrMnB stainless overlay and 0Cr13Ni5Mo stainless steel in 0.5M NaCl and 0.5M HCl solutions. Tribology International, **41**(12): 1181-1189.

Zheng, Y. G., S. Z. Luo, and W. Ke, 2007. Effect of passivity on electrochemical corrosion behavior of alloys during cavitation in aqueous solutions, Wear, **262**(11-12): 1308-1314.

Zhu, Z. F., 1995. Corrosion resistance and application of non-ferrous alloys, Beijing Chemical Industry Press, Beijing, 99-105,107-109.

Zhuecheva, A., W. Sha, S. Malinov and A. Long, 2005. Enhancing the microstructure and properties of titanium alloys through nitriding and other surface engineering methods. Surface and Coatings Technology, **200**(7): 2192-2207.

Subject Index

Terminologies

Some terminologies relating to materials engineering used in this book are explained below:

Intermetallic compound | A compound that is composed of metal atom or ion and non-metal atom or ion, having a characteristic crystal structure, and a metallic bonding.

Covalent compound | A compound formed from the covalent bonding of two or more elements.

Ionic compound | A compound formed from the chemical bonding of element with opposite charges.

Solid solution | A phase formed by adding atoms of one element (solute) to the crystalline lattice of another element (solvent). If the solute metal atom enters one of the interstices between the solvent atoms, it is an interstitial solid solution. If the solute metal atoms replace the solvent metal atoms in the crystal lattice, it is a substitutional solid solution. When the solute and solvent atoms have infinitely mutual solubility, it is called all proportional solid solution; otherwise it is limit solid solution.

Standard electrode potential | Electric potential of an electrode reaction relative to a standard hydrogen electrode (SHE) using an effective concentration of 1 $mol \cdot dm^{-3}$ for all species in solution at a standard state (temperature of 25°C and pressure of 1 atm.).

Corrosion potential | The voltage difference between a metal immersed in a given medium and a reference electrode. It is also called open circuit potential or mixed corrosion potential.

Corrosion current density	The current density at the corrosion potential. It can be used to calculate the corrosion rate of the metal.
Pitting potential	The electrochemical potential in a given enviroment above which pits can form in a metal surface.
Maintaining passive potential	The minimum potential required to maintain the thickness of film in the passive region.
Maintaining passive current density	The minimum current density required to maintain the thickness of film in the passive region.
Transpassive potential	The potential that marks the end of the passive potential region and the transition from passive to tranpassive behavior.
Galvanic series	A table showing semi-metal and metal nobility. It is also known as electropotential series. It can give a qualitative guide for the selection of material and structure design. In galvanic series, the closer materials are, the lower the galvanic corrosion risk.
Fracture toughness	A property which describes the ability of a material containing a crack to resist fracture (K_I). $$K_I = 0.004985\left(\frac{E}{HV}\right)^{\frac{1}{2}}\frac{P}{C^{\frac{3}{2}}}$$ where P is the load (N), C is the length of crack (mm), E is Young's modulus (GPa) and HV is microhardness (GPa).
Anodic passivation	A method to decrease the corrosion of a metal by making it an anode in an electrochemical cell and controlling its potential in a zone where the metal is in passive state.

About the authors:

Cui Lin holds a bachelor's degree in Chemical Engineering and a master's degree in Materials Processing Engineering from Nanchang Hangkong University. She then earned a Ph.D. degree in Materials Engineering from the University of Science and Technology Beijing, China in 2004. In the same year, she joined Nanchang Hangkong University as a faculty member and was promoted to Professor in 2009. Between 2010 and 2016, she had opportunities to learn, teach and conduct research in the materials engineering and mineral resource engineering at Dalhousie University.

She has 10 years of teaching and research experience in the corrosion and protection engineering fields. She has taught a number of undergraduate and graduate courses in materials engineering and mining engineering, and supervised more than 10 graduate students. Over the years, she has completed many research projects and published over 30 papers in referred journals and conference proceedings and 2 books plus 2 patents. Her research work involves atmospheric corrosion, cavitation erosion, chemical etching of non-ferrous metals and refractory alloys, corrosion monitoring and electroless plating, mine rock mechanics.

Meifeng Wang was awarded a Ph.D. by the University of Science and Technology Beijing, China in 2009 for his work on pitting corrosion of metals. Currently he is an Associate Professor in the Department of Corrosion and Protection at Nanchang Hangkong University. His main research interests include chemical treatment on metal surfaces and corrosion mechanisms of metals.